AF608897

Debottlenecking the Evaporator in Water-Based Adsorption Chillers

Verdampferentwicklung zur Überwindung von Engpässen in wasserbasierten Adsorptionskältemaschinen

Von der Fakultät für Maschinenwesen der
Rheinisch-Westfälischen Technischen Hochschule Aachen
zur Erlangung des akademischen Grades eines
Doktors der Ingenieurwissenschaften
genehmigte Dissertation

vorgelegt von

Jan Michael Seiler

Berichter: Univ.-Prof. Dr.-Ing. André Bardow
apl. Prof. Dr.-Ing. Hans-Jürgen Koß
Univ.-Prof. Dr.-Ing. Stephan Kabelac

Tag der mündlichen Prüfung: 12. März 2021

Aachener Beiträge zur Technischen Thermodynamik Band 32

Jan Michael Seiler
Debottlenecking the Evaporator in Water-Based Adsorption Chillers
Verdampferentwicklung zur Überwindung von Engpässen
in wasserbasierten Adsorptionskältemaschinen

ISBN: 978-3-95886-407-8

Bibliografische Information der Deutschen Bibliothek
Die Deutsche Bibliothek verzeichnet diese Publikation in der Deutschen Nationalbibliografie; detaillierte bibliografische Daten sind im Internet über http://dnb.ddb.de abrufbar.

Herstellung & Vertrieb:

1. Auflage 2021

Süsterfeldstr. 83, 52072 Aachen
Tel. 0241 / 87 34 34 00
www.Verlag-Mainz.de

ISSN: 2198-4832

Satz: nach Druckvorlage des Autors
Umschlaggestaltung: Druckerei Mainz

printed in Germany
D82 (Diss. RWTH Aachen University, 2021)

Danksagung

Die vorliegende Arbeit entstand im Rahmen meiner Tätigkeit als wissenschaftlicher Mitarbeiter am Lehrstuhl für Technische Thermodynamik der Rheinisch-Westfälischen Technischen Hochschule Aachen. An dieser Stelle möchte ich den Menschen meinen Dank aussprechen, die diese Arbeit erst ermöglichten.

Zuerst möchte ich mich bei Univ.-Prof. Dr.-Ing. André Bardow für das mir entgegengebrachte Vertrauen, die gewährten Freiheiten in der Forschung und die erfolgreiche Zusammenarbeit bedanken. Insbesondere seine Art und Weise wissenschaftliche Inhalte zu kommunizieren, empfinde ich als sehr inspirierend. Sowohl apl. Prof. Dr.-Ing. Hans-Jürgen Koß als auch Univ.-Prof. Dr.-Ing. Stephan Kabelac danke ich ganz herzlich für die Übernahme der Koreferate. Bei Univ.-Prof. Dr.-Ing. Peter Jeschke möchte ich mich für die Übernahme des Prüfungsvorsitz und die angenehme Leitung der Prüfung bedanken.

Mein herzlicher Dank gilt allen aktuellen und ehemaligen Kolleginnen und Kollegen am Lehrstuhl: Ihr macht den LTT zu etwas Besonderem! Ganz besonderer Dank geht an die Mitglieder der Arbeitsgruppe Sorptionstechnologie: Franz Lanzerath danke ich für die Einführung in die Welt der Sorptionstechnologie während meiner Masterarbeit und dafür, dass er die Grundlage für meine Arbeit gelegt hat; Heike Schreiber möchte ich für die vielen konstruktiven Diskussionen und für ihre Fürsorge in wichtigen Momenten danken; Meltem Erdoğan danke ich für die Vorarbeiten zum Thema Verdampfung und unsere lebhaften Diskussionen; Uwe Bau danke ich ganz besonders für die gemeinsamen Arbeiten, für das Feedback zu dieser Arbeit und die Freundschaft, die uns verbindet; Stefan Graf möchte ich für die langjährige Freundschaft und die Unterstützung auf dem gemeinsamen Weg danken; Andrej Gibelhaus danke ich für die Diskussionen im Büro und die gemeinsame Zeit jenseits der Arbeit; Dominik Tillmanns danke ich für die Leichtigkeit und Freude, die er in die Arbeitsgruppe bringt; der neuen Generation bestehend aus Mirko Engelpracht, Marten Entrup, Matthias Henninger und Patrik Postweiler danke ich für die tolle Zusammenarbeit, die gemeinsamen Boßeltouren und für die Fortführung der spannenden Forschung in der Arbeitsgruppe.

Diese Arbeit ist nur durch das Engagement, die Hilfe und die Unterstützung zahlreicher Studierenden möglich geworden, denen ich hier besonderen Dank aussprechen möchte: Jonas Hackmann, Sebastian Jörres, Fabian Fröde, Matthias Vogelsang, Christian Zibunas, Sophia Schröer, Nils Baumgärtner, Julian Lützow, Niklas Houba, Dominic Dittmer, Dominik Kemetmüller, Stephan Göbel, Dennis Krakau, Dennis Derwein und Björn Cottmann.

Mein ganz besonderer Dank gilt den Mitarbeitenden aus Werkstätten und Labor, ohne die kein Experiment hätte aufgebaut und durchgeführt werden können. Daher danke ich Thomas Bungert, Djuro Dragas, Edgar Brauers, Bernhard Müller, Hartmut de Vries, Ton Ritterbex und Sonja Kröngen für ihren unglaublich wertvollen Erfahrungsschatz und die tatkräftige Unterstützung bei allen experimentellen Arbeiten. Ebenfalls sehr dankbar bin ich für die organisatorische Unterstützung und Hilfe durch Iris Wallraven, Katrin Roßbruch und Eva Frach sowie den IT-Support durch Gregor Migas.

Abschließend möchte ich mich bei meinen Freunden und meiner Familie für die Begleitung und Unterstützung während der Promotion bedanken. Meinen Eltern danke ich neben der immerwährenden Unterstützung insbesondere für die Begeisterung für die Wissenschaft, die sie mir mit auf den Weg gegeben haben. Großer Dank geht an meine Schwiegereltern und meine Familie, deren Hilfe bei der Betreuung unserer Kinder während Corona-Zeiten unglaublich wertvoll war für das Verfassen dieser Arbeit. Meiner Freundin Caroline Marks und unseren Töchtern Carlotta und Leni gilt mit Abstand mein größter Dank: Caroline, Du hast mir die Freiräume gegeben die Promotion abzuschließen und mir dabei immer wieder Zuversicht geschenkt. Du und unsere beiden Töchter zeigen mir jeden Tag aufs Neue was wirklich wichtig ist.

Zürich, im März 2021 *Jan Seiler*

'Man muss das Unmögliche versuchen, um das Mögliche zu erreichen.'

Hermann Hesse, (1877 – 1962)

Contents

Appendices

Kurzfassung

Der weltweite Kältebedarf steigt kontinuierlich. Um daraus resultierende Umweltauswirkungen zu verringern, ist nachhaltige Kältetechnik notwendig. Adsorptionskältemaschinen können Kälte nachhaltig bereitstellen, da sie sowohl umweltfreundliche Abwärme oder Sonnenenergie nutzen als auch mit dem natürlichen Kältemittel Wasser betrieben werden können.

Wasser vereint als Kältemittel viele Vorteile: Es hat kein Treibhauspotential, ist absolut sicher, günstig, gut verfügbar und hat eine hohe Verdampfungsenthalpie. Allerdings gibt es auch technische Hürden bei der Nutzung als Kältemittel: Einerseits ist die effiziente Niederdruckverdampfung von Wasser anspruchsvoll. Anderseits limitiert der Gefrierpunkt von Wasser den Anwendungsbereich auf Temperaturen über 0 °C, weshalb viele Kühlanwendungen nicht bedient werden können. Beide Nachteile des Kältemittels Wasser werden in dieser Arbeit adressiert: (I) Die Dünnfilmverdampfung im Niederdruck ermöglicht einen effizienten Wärmeübergang und wird in dieser Arbeit experimentell untersucht. (II) Die Limitierung auf Temperaturen über den Gefrierpunkt kann aufgehoben werden, indem ein Frostschutzmittel in den Verdampfer gegeben wird. Experimente mit einer Adsorptionskältemaschine zeigen die Machbarkeit.

Die Experimente zur Niederdruckverdampfung von Wasser zielen darauf ab, kapillargestützte Verdampfer besser zu verstehen und dadurch den Wärmeübergang beim Phasenwechsel steigern zu können. Zuerst wird der Versuchsaufbau validiert, indem identische Experimente an zwei Versuchsaufbauten verglichen werden. Der Vergleich liefert zusätzlich Einsichten zu grundsätzlichen Einflussfaktoren auf die Dünnfilmverdampfung im Niederdruck. Ebenfalls validiert wird eine spezielle Versuchsdurchführung mit kontinuierlich sinkenden Füllständen, bei der Wärmeübertrager anhand eines einzelnen Experiments schnell bewertet werden können. Der validierte Versuchsaufbau wird genutzt, um den Wärmeübergang bei der Dünnfilmverdampfung von 7 kapillaraktiven Beschichtungen zu charakterisieren. Durch Korrelation der Oberflächeneigenschaften der unterschiedlichen Beschichtungen zum Wärmeübergang können vorteilhafte Oberflächeneigenschaften identifiziert werden.

Der Betrieb einer wasserbasierten Adsorptionskältemaschine unterhalb von 0 °C wird durch die Zugabe von Ethylenglykol ermöglicht. Die Auswirkungen des Frostschutzmittels werden bei unterschiedlichen Betriebsbedingungen im Experiment analysiert und bewertet. Sich ergebende Herausforderungen für eine kommerzielle Nutzung werden identifiziert und Möglichkeiten zur Prozessverbesserung diskutiert.

Die Arbeit liefert somit einen Beitrag, um die hervorragenden Eigenschaften des natürlichen Kältemittels Wasser für die Kältebereitstellung besser nutzbar zu machen.

Abstract

Global cooling demand continues to grow. To mitigate the resulting environmental impact, cooling solutions need to be sustainable. Adsorption chillers can provide sustainable cooling as they can employ both environmentally friendly driving forces such as solar or waste heat and the natural refrigerant water.

Water offers many advantages as refrigerant: its global warming potential is zero, it is absolutely safe, cheap, broadly available and has a high enthalpy of vaporisation. However, there are also challenges to employ water as refrigerant in cooling applications. On the one hand, efficient sub-atmospheric evaporation of water is challenging. On the other hand, the freezing point of water limits its operational range to temperatures above 0 °C, thereby excluding many cooling applications. Both bottlenecks of the natural refrigerant water are addressed experimentally in this thesis: (I) Sub-atmospheric, thin-film evaporation of water enables efficient heat transfer and is investigated using capillary-assisted heat exchangers. (II) The limitation towards temperatures above 0 °C is overcome by adding the anti-freezing agent ethylene glycol to the evaporator of an adsorption chiller, and the feasibility of operation is demonstrated.

The experiments regarding sub-atmospheric evaporation aim at understanding capillary-assisted, thin-film evaporators to improve their heat transfer performance. First, the employed experimental setup is validated by comparing results of identical experiments using two setups. The comparison allows to identify influencing factors for experimental studies of sub-atmospheric, thin-film evaporation in general. Experiments with continuously decreasing filling levels are experimentally assessed since the procedure allows to analyse all filling levels in a single experiment. Next, the validated experimental setup and procedure are employed to analyse the evaporation performance of 7 coated tubes that create capillary action for thin-film evaporation on their surfaces. By correlating surface properties to evaporation performance for different coatings and experimental conditions, favourable surface properties and testing conditions are derived.

Water-based adsorption cooling below 0 °C is investigated in a one-bed adsorption chiller. The impact of adding ethylene glycol as an anti-freezing agent to the evaporator is assessed at different working conditions and the resulting process performance is evaluated. Challenges regarding successful implementation in commercial systems and future integration of thin-film evaporation for process improvement are discussed.

Thus, both discussed bottlenecks are overcome, thereby enabling even further exploration of the outstanding properties of the natural refrigerant water.

List of Figures

List of Tables

Notation

Latin symbols

a	interval	$[a]$
A	reference area for heat transfer	m^2
A_{eff}	effective surface area for heat transfer	m^2
A-C	names of repetitions of experiments in Chapter 4	
A-H	names of coatings examined in Chapter 5	
A	calibration sensor A	
B	calibration sensor B	
c	specific heat capacity	$J\,kg^{-1}\,K^{-1}$
d	diameter	m
d_o	outer diameter of tube (from fin tip to fin tip)	m
f_{rel}	relative filling level of evaporator (cf. Figure 3.6)	
f	function	$[f]$
g	gravity of earth	$9.81\,m\,s^{-2}$
h	specific enthalpy	$J\,kg^{-1}$
Δh_{ads}	specific enthalpy of adsorption	$J\,kg^{-1}$
h_{inner}	inner heat transfer coefficient	$W\,m^{-2}\,K^{-1}$
h_{outer}	outer heat transfer coefficient	$W\,m^{-2}\,K^{-1}$
H	height	m
i	control variable	
k	coverage factor for expanded uncertainty according to GUM nomenclature	
l	length of surface contour	µm
L	length of one tube	m
m	mass	kg
$\dot{m}$	mass flow	$kg\,s^{-1}$
n	number	

p	pressure	mbar
P	probability	
$\dot{q}$	heat flux	$\mathrm{J\,s^{-1}\,m^{-2} = W\,m^{-2}}$
Q	transferred heat	kJ
$\dot{Q}$	heat flow	$\mathrm{J\,s^{-1} = W}$
r	Pearson correlation coefficient	
Re	Reynolds number	
R_z	surface roughness (cf. Table 2.1)	µm
s	valley-to-peak distance	µm
s_x	experimental standard deviation of measured quantity x	$[x]$
t	time	s
t_P	Student t-factor	
T	temperature	°C or K
ΔT_ln	logarithmic mean temperature difference	K
u_x	standard measurement uncertainty of quantity x	$[x]$
$u_{\mathrm{c},f(x_i)}$	combined standard uncertainty of the function f	$[f]$
U	overall heat transfer coefficient	$\mathrm{kW\,m^{-2}\,K^{-1}}$
UA	overall heat transfer coefficient U multiplied by heat transfer area A	$\mathrm{kW\,K^{-1}}$
v	velocity	$\mathrm{m\,s^{-1}}$
$\dot{V}$	volume flow	$\mathrm{L\,min^{-1}}$
w	loading of adsorbent	$\mathrm{kg_{refrigerant}\,kg_{sorbent}^{-1}}$
w_add	mass fraction of additive	$\mathrm{kg_{additive}\,kg_{mixture}^{-1}}$
x	quantity	$[x]$
$\bar{x}$	arithmetic mean	$[x]$
y	variable	$[y]$
z	vertical position	m

Greek symbols

δ	thickness of the film	µm
δ_c	thickness of the coating	µm
δ_λ	thickness of the plate	m
Δ	difference	

Δh_{ads}	specific enthalpy of adsorption	$\mathrm{J\,kg^{-1}}$
η	dynamic viscosity	$\mathrm{Pa\,s}$
θ	contact angle	$^{\circ}$
κ	interface curvature	$\mathrm{m^{-1}}$
λ	heat conductance	$\mathrm{W\,m^{-1}\,K^{-1}}$
μ	chemical potential	$\mathrm{J\,mol^{-1}}$
ν	degrees of freedom	
ϱ	density	$\mathrm{kg\,m^{-3}}$
σ	interfacial tension	$\mathrm{N\,m^{-1}}$

Subscripts

0	start or beginning
1	end
a	advancing
ad	adsorbate
add	additive
ads	adsorption
A	adsorber
Ads	adsorption phase
c	coating
c	combined
cap	capillary
cond	condenser
cycle	full adsorption cycle consisting of of 4 phases according to Figure 2.2
C	condenser
dis	disjoining pressure
Des	desorption phase
eff	effective
E	evaporator
hx-fluid	heat exchanger fluid
in	inlet
IC	isosteric cooling phase
IH	isosteric heating phase
l	liquid
l-v	liquid-vapour
ln	logarithmic

max	maximum
min	minimum
o	outer
out	outlet
r	receding
rel	relative
sat	saturation
sor	sorbent
v	vapour
w	wall
z	see Latin letter R_z

Superscripts

$\bar{x}$	arithmetic mean (cf. Equation (C.1) for details)
$\overline{X}$	weighted average (cf. Equation (5.3) for details)

Abbreviations and Acronyms

a	advancing
ad	adsorbate
add	additive
ads	adsorption
A	adsorber
Ads	adsorption phase
ASHRAE	American Society of Heating, Refrigeration and Air-Conditioning Engineers
c	coating
c	combined
cap	capillary
cond	condenser
C	condenser
Ca	capillary number
Cu	copper
CAD	computer aided design

CHF	critical heat flux
COP	coefficient of performance
CP	coating layer porosity
CV	coefficient of variation
dis	disjoining pressure
DAkkS	Deutsche Akkreditierungsstelle GmbH: German national accreditation body
Des	desorption phase
eff	effective
E	evaporator
fpi	fins per inch
FC-72	flurocarbon perfluorohexane (C_6F_{14}) often used as liquid electronic coolant
GUM	guide to the expression of uncertainty in measurement
GWP	global warming potential
highDF	high driving force
hx	heat exchanger
in	inlet
IAPWS	International Association for the Properties of Water and Steam
IC	isosteric cooling phase
IH	isosteric heating phase
ISE	Fraunhofer Institute for Solar Energy systems
ISO-KF	Normung für vakuumfeste Kleinflansche: vacuum-tight ISO standard quick release flange
l	liquid
l-v	liquid-vapour
lowDF	low driving force
ln	logarithmic
LED	light-emitting diode
LT	coating layer thickness
max	maximum
min	minimum
NIST	National Institute of Standards and Technology
o	outer
out	outlet
ONB	onset of nucleate boiling
Pt100	platinum resistance thermometer with a resistance of $100\,\Omega$ at $0\,^{\circ}C$
PMMA	Poly(methyl methacrylate)

r	receding
rel	relative
rgb	red, green and blue
Re	Reynolds number
RefProp	NIST Reference Fluid Thermodynamic and Transport Properties database
RWTH	Rheinisch–Westfälische Technische Hochschule Aachen
R-718	refrigerant water (ASHRAE Number)
sat	saturation
sor	sorbent
SCP	specific cooling power
SE	surface extension
SI	supplementary information
v	vapour
w	wall

Chapter 1

Introduction

In response to the global threat of climate change, the parties to the United Nations Framework Convention on Climate Change adopted the Paris Agreement in 2015 to "holding the increase in the global average temperature to well below 2 °C above pre-industrial levels and pursuing efforts to limit the temperature increase to 1.5 °C above pre-industrial levels" (United Nations, 2015).

Although many countries have ratified the Paris Agreement since, in 2018, a special report by the Intergovernmental Panel on Climate Change showed that "recent trends in emissions and the level of international ambition indicated by nationally determined contributions, within the Paris Agreement, deviate from a track consistent with limiting warming to well below 2 °C". Additionally, the authors state that "without increased and urgent mitigation ambition in the coming years, leading to a sharp decline in greenhouse gas emissions by 2030, global warming will surpass 1.5 °C in the following decades, leading to irreversible loss of the most fragile ecosystems, and crisis after crisis for the most vulnerable people and societies" (Masson-Delmotte et al., 2018). Therefore, urgent actions need to be taken to combat climate change and reduce greenhouse gas emissions.

In 2014, 8 % of greenhouse gas emissions worldwide were caused by refrigeration (International Institute of Refrigeration et al., 2017). According to the the International Energy Agency (2018a), the demand for cooling continues to increase and will be the second-largest driver for growth in electricity demand (behind electrical motors) until 2040. Consequently, with currently used technologies, refrigeration will also increase greenhouse gas emissions. Greenhouse gas emissions caused by refrigeration can be divided into indirect and direct emissions. About 63 % are indirect emissions attributed to production and transport of the energy required to operate the refrigeration systems. The remaining 37 % are direct emissions, e. g. caused by leakage of the employed refrigerants (International Institute of Refrigeration et al., 2017).

Both types of emissions can be mitigated by water-based adsorption chillers: adsorption chillers can use solar or waste heat as driving energy instead of electricity consumed by conventional compression chillers (Meunier, 2013). Furthermore, adsorption chillers allow employing the natural refrigerant water. Besides having a global warming potential of zero, water also offers the additional advantages of non-toxicity, broad availability and low costs (Fernandes et al., 2014).

Despite all the advantages of water-based adsorption chillers, current market penetration is negligible and "solar thermal cooling systems using sorption chillers are to date relatively niche technologies" (International Energy Agency, 2018b). The main reasons that hinder the success of adsorption refrigeration systems are high investment costs as well as poor heat and mass transfer (Li et al., 2015) and, thus, poor power density (Wang et al., 2014a). To increase power density, not only the adsorber design but also the evaporator design is important (Lanzerath et al., 2015; San and Tsai, 2014).

Evaporator design is challenging especially for the natural refrigerant water since the heat transfer process is inefficient. Furthermore, current water-based adsorption chillers cannot provide all cooling services because employing pure water as refrigerant is not feasible at temperatures below 0 °C due to freezing (Freni et al., 2015; Wang et al., 2009, 2014b). Thus, the natural refrigerant water suffers from two bottlenecks: (I) efficient, sub-atmospheric evaporation is challenging and (II) the operational range of water is limited to temperatures above the freezing point.

This thesis aims at overcoming both bottlenecks in the evaporator of an adsorption chiller: efficient, sub-atmospheric evaporation of the natural refrigerant water is enabled by thin-film evaporation and is experimentally investigated. The limitation of water towards lower temperatures is overcome by adding the anti-freezing agent ethylene glycol to the evaporator, rendering water-based adsorption cooling below 0 °C possible.

Structure of the thesis

This thesis focuses on the evaporator of water-based adsorption chillers and addresses two bottlenecks of the natural refrigerant water: (I) efficient, sub-atmospheric evaporation and (II) operation at temperatures below the freezing point.

Chapter 2 provides an overview of the fundamentals for adsorption cooling in general and sub-atmospheric evaporation of water in detail. Starting from the working principle of adsorption chillers and their traditional key performance indicators, advantages and limitations are identified for water-based adsorption chillers. Next, the state-of-the-art for evaporators in adsorption chillers is reviewed, and specifics are derived for an ideal evaporator. For pool boiling, fundamentals of evaporation and boiling regimes are presented and the boiling regime for thin-film evaporation investigated in this thesis is located on the Nukiyama curve. Challenges of sub-atmospheric, thin-film evaporation are explained and the analogy to evaporation in heat pipes is drawn. Associated concepts regarding interactions between refrigerants and solids, which are necessary for the present analysis, are also introduced. This includes capillary action to maintain thin refrigerant films on the heat exchanger's surfaces. Finally, theoretical analysis for evaporation from an extended meniscus (e. g. in a capillary) as a possible evaporation mechanism is summarised. Chapter 2 concludes with both a summary of the presented fundamentals and the contributions of this thesis.

Chapter 3 introduces the two experimental setups employed in this thesis. The adsorption experiments were performed in a one-bed adsorption chiller setup and the evaporation experiments were conducted in a custom-built evaporator setup to determine heat transfer coefficients. The descriptions of both setups follow the path of the gathered experimental information: from the physical experiment, through sensors for data acquisition, to data reduction. The chapter concludes with an analysis of the uncertainty in measurement of the employed sensors.

In Chapter 4, the experimental setup for the evaporation experiments is validated. Comparing the results obtained from identical experiments at another experimental setup allows to disclose prerequisites for obtaining comparable results for sub-atmospheric evaporation in general. Furthermore, different experimental procedures are compared to establish the validity of the employed experimental procedure as well.

In Chapter 5, the evaporation performance is investigated for different coatings that enable capillary-assisted, thin-film evaporation. Seven coatings and a plain tube are analysed at different experimental conditions in an automated setup. The results reveal advantages of certain structural properties. Subsequently, the surface properties are correlated to evaporation performance, thereby allowing to derive a guideline for future coating development.

In Chapter 6, the feasibility of water-based adsorption cooling below 0 °C is experimentally evaluated. The anti-freezing agent ethylene glycol is added to the evaporator to prevent freezing of the refrigerant water. Experiments are conducted with varying mass fractions of the anti-freezing agent and different cooling temperatures to analyse their impact on adsorption chiller performance.

Chapter 7 provides a summary of the thesis, contains the major conclusions and gives an outlook for future work.

Chapter 2

State-of-the-art and limitations of evaporators for water-based adsorption chillers

This chapter gives a brief overview of fundamentals relevant for this thesis. Following the two main investigated subjects, (I) efficient, sub-atmospheric evaporation of water and (II) operation of water-based adsorption chillers below 0 °C, the presented fundamentals are divided into the following two sections. Section 2.1 introduces adsorption chillers and emphasises challenges for the natural refrigerant water. As both main investigated subjects are related to the evaporator of adsorption chillers, a special focus lies on the evaporator. Section 2.2 provides an overview of evaporation in general and of sub-atmospheric evaporation of water in detail.

2.1 Adsorption chillers

Adsorption chillers are thermally driven: they use heat (e. g. solar or waste heat) to supply cooling power. Thus, adsorption chillers can substitute commonly used mechanical compression chillers and avoid greenhouse gas emissions in two ways: (I) emissions connected to the production of electricity for substituted mechanical refrigeration and (II) emissions due to incidental release of refrigerants with a high global warming potential (GWP), since adsorption chillers allow to use low GWP natural refrigerants – preferably water.

Although the focus of this work lies on refrigeration, and therefore on adsorption chillers, the identified challenges and solutions also apply to other adsorption energy systems like thermal storage (Aydin et al., 2015) or heat pump systems (Meunier, 2013): these systems also benefit from improving the evaporator performance, and expanding the operational temperature range for water adsorption applications below 0 °C is also relevant, for example, for domestic air-source heat pumps in freezing winter conditions.

This section gives a brief overview from the working principle of adsorption to the performance indicators of adsorption chillers (Section 2.1.1), the characteristic advantages and challenges of adsorption chillers using the natural refrigerant water (Section 2.1.2) and currently used evaporators in adsorption chillers (Section 2.1.3).

2.1.1 Adsorption: from working principle to cycle performance

Physical adsorption describes the process of fluid molecules attaching to the surface of a solid due to van-der-Waals forces and electrostatic interactions. The nomenclature of the involved phases is defined in Figure 2.1: during **adsorption**, the fluid (**adsorptive**; in this thesis always assumed to be in the vapour phase prior to adsorption respectively after desorption) is attached to the solid (**adsorbent**) and is called **adsorbate** in the adsorbed state. The process is fully reversible: **desorption** is the process of molecules detaching from the surface of the solid. The combination of adsorptive and adsorbent is referred to as the working pair.

Thermodynamics of adsorption

For thermally-driven adsorption chillers, balancing energies during the adsorption process is important: since the energy level of the adsorbate is lower than the energy level of the adsorptive, adsorption is an exothermic and desorption an endothermic process. Thus, during adsorption, the specific enthalpy of adsorption Δh_{ads} given by

$$\Delta h_{\mathrm{ads}} = h_{\mathrm{v}} - h_{\mathrm{ad}} \tag{2.1}$$

is released. Here, h_{v} is the specific enthalpy of the adsorptive in the vapour phase and h_{ad} is the specific enthalpy of the adsorbate in the adsorbed state. The specific enthalpy of the adsorbate h_{ad} depends on the temperature T of the adsorbate and the amount of adsorbate already present in the adsorbent, described by the loading w. The loading w of the adsorbent relates the mass of adsorbate m_{ad} to the mass of dry adsorbent m_{sor}:

$$w = \frac{m_{\mathrm{ad}}}{m_{\mathrm{sor}}}\,. \tag{2.2}$$

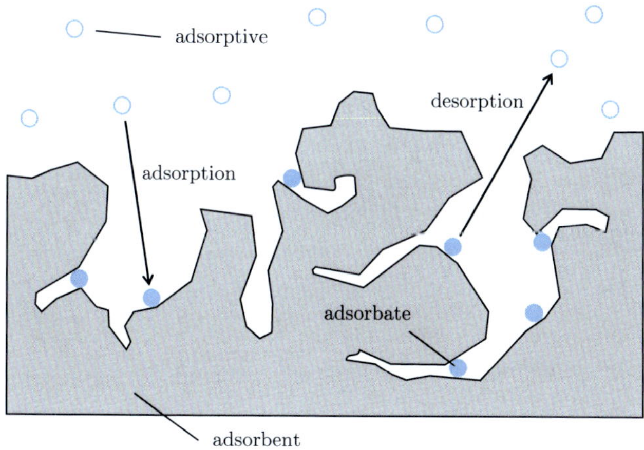

Figure 2.1: Definition of the nomenclature for the adsorption process. The figure is based on Bathen and Breitbach (2001); Schnabel (2009) and Graf (2018). In their German publication, Bathen and Breitbach (2001) further differentiate between the molecules of the fluid in the adsorbed state ("Adsorpt") and the complex of adsorbed fluid and adsorbent ("Adsorbat"), yet this differentiation is not relevant in this thesis.

Adsorption increases the loading while desorption decreases it. The loading changes as long as the adsorbate is not in thermodynamic equilibrium with the surrounding vapour phase of the adsorptive given by the inequality of their chemical potentials (Ruthven, 1984):

$$\mu_{\mathrm{ad}} \neq \mu_{\mathrm{v}} \,. \tag{2.3}$$

Once in equilibrium, the loading w for pure component adsorption can be described by two independent variables. Usually, the chosen independent variables are temperature T and pressure p. Changes in temperature or pressure trigger ad- or desorption and vice versa: a higher adsorbent temperature results in desorption. The desorbed molecules are released into the vapour phase and increase the pressure in a closed system. Contrarily, decreasing the temperature of the adsorbent leads to adsorption and, thus, a decrease of pressure in the vapour phase.

The decrease in pressure can be used to create the driving force for evaporation from a liquid phase of the adsorptive. Decreasing the pressure in the vapour phase affects the vapour–liquid equilibrium and results in evaporation from a liquid phase to replace the missing molecules in the vapour phase. The evaporation occurs at the saturation temperature of the adsorptive, which is lower than the temperature of the adsorbent in equilibrium. Thus, the enthalpy of vaporisation can be used to supply cooling power at the lower saturation temperature. The described process is used in adsorption chillers to provide cooling power. The resulting thermodynamic cycle is presented next.

Adsorption chiller cycle

The components of an adsorption chiller are very similar to the components of the commonly used compression chillers: only the compressor is replaced by an adsorber. Thus, adsorption chillers consist of the following components: evaporator (E), adsorber (A) and condenser (C) (the colours are used for the components throughout the thesis: cf. Figure 2.2). Instead of consuming electricity to power the compression, adsorption chillers employ thermal compression and use heat to drive the process. During adsorption, the enthalpy of adsorption (Equation (2.1)) needs to be removed as a heat flow from the adsorber at a low-temperature level, while desorption requires heat for regeneration of the adsorber at a high-temperature level.

Overall, there are 4 temperature levels in the components of an adsorption chiller:

- Evaporation in evaporator at **low** temperature
- Adsorption in adsorber at **medium** temperature
- Desorption in adsorber at **high** temperature
- Condensation in condenser at **medium** temperature

Often, the medium temperatures for adsorption and condensation are chosen to be identical, which is also the case in this thesis.

In contrast to compression chillers, adsorption chillers work discontinuously, and the cycle for a 1-bed chiller can be divided into 4 phases. These phases are visualised in a process flow sheet in Figure 2.2 showing the necessary components, valves and heat/mass flows in operation. Additionally, for the ideal cycle, the thermodynamic states of the adsorbent and the heats necessary for the change of states are displayed in an Arrhenius diagram ($\ln(p)$ over $-1/T$) in Figure 2.3.

Adsorption phase (Ads) ① → ②

During the adsorption phase, the valve between the adsorber and the evaporator is open. Refrigerant in the vapour phase of the evaporator at pressure p_E is adsorbed in the adsorber, thus creating the mass flow $\dot{m}$ from the evaporator

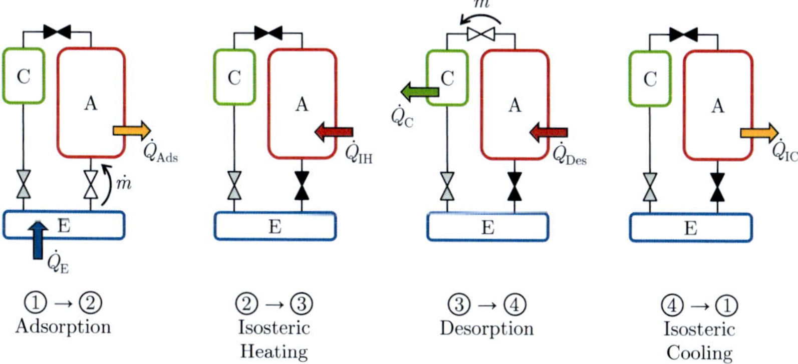

Figure 2.2: Simple 1-bed adsorption chiller cycle: the 3 main components adsorber (A), evaporator (E), and condenser (C) are connected by valves; states of the valves are given by colour (black: closed, white: open, grey: for liquid reflux); mass flows and heat flows are shown for the 4 phases: adsorption, isosteric heating, desorption and isosteric cooling. This figure is based on Lanzerath (2014).

to the adsorber and causing evaporation of refrigerant in the evaporator. The resulting heat flow $\dot{Q}_E$ into the evaporator is the main purpose of an adsorption chiller: cooling power.

During adsorption, the loading of the adsorbent w increases and the enthalpy of adsorption Δh_{ads} is released. The heat flow $\dot{Q}_{Ads}$ cools the adsorber to the medium temperature T_{Ads}, which is reached at the end of the adsorption phase when the adsorbent is in equilibrium at the maximum loading of the cycle w_{max}.

Isosteric Heating phase (IH) ② → ③

To reverse the adsorption process, the adsorber is heated. The first part of the heating process takes place at constant loading, i. e. isosteric. In the isosteric heating phase, all valves are closed and the heat flow $\dot{Q}_{IH}$ increases the adsorber's temperature until the pressure $p_C(T_C)$ of the condenser is reached. The corresponding temperature of the adsorber at the end of the isosteric heating phase is T_{IH}.

Desorption phase (Des) ③ → ④

The desorption process starts, once the valve between the adsorber and the condenser is opened and the refrigerant mass flow $\dot{m}$ streams from the adsorber

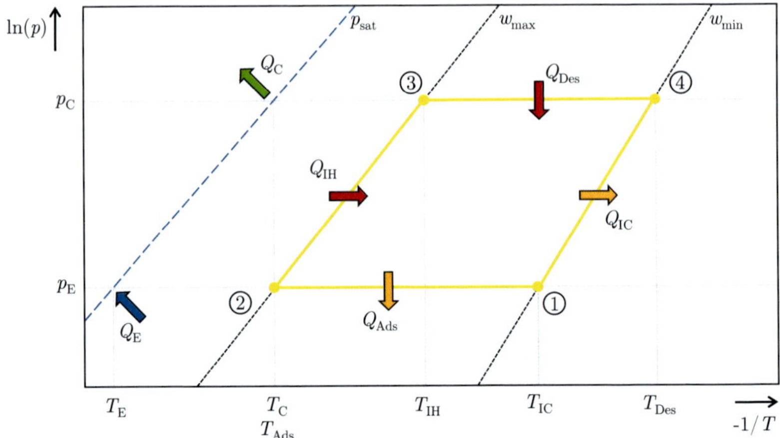

Figure 2.3: Adsorbent states ① - ④ of simple 1-bed-adsorption-chiller cycle plotted in an Arrhenius diagram ($\ln(p)$ over $-1/T$) where p is the pressure of the adsorptive and T the temperature of the adsorbent in Kelvin. The heats Q_i necessary for the change of states and transferred during the phases are also included. Shown as dashed lines in black are isosters, i. e. states with the same loading w, and the saturation pressure p_{sat} of water in blue. The adsorbent is cycled between the minimum loading w_{min} and maximum loading w_{max}. The adsorbent's temperature changes from T_{Ads} to T_{Des}. Ad- resp. desorption takes place at evaporator pressure $p_E(T_E)$ / condenser pressure $p_C(T_C)$. This figure is based on Núñez (2001).

to the condenser. Refrigerant condenses in the condenser at pressure p_C releasing the enthalpy of vaporisation which is removed from the condenser by the heat flow $\dot{Q}_C$. The heat flow $\dot{Q}_{Des}$ drives the desorption by supplying the enthalpy of adsorption Δh_{ads} to the adsorbent and increasing the adsorbent's temperature. The desorption phase ends when the adsorbent is in equilibrium at temperature T_{Des} and minimal loading of the cycle w_{min}.

Isosteric Cooling phase (IC) ④ $\rightarrow$ ①

To close the adsorption cycle, the dried, hot adsorbent needs to be cooled down for adsorption again. This cooling process is isosteric. Thus, all valves are closed and, during isosteric cooling, the heat flow $\dot{Q}_{IC}$ leaves the adsorbent until the temperature T_{IC} of the adsorbent and the pressure $p_E(T_E)$ of the evaporator are reached.

Liquid reflux

Liquid reflux from the condenser to the evaporator (grey valve in Figure 2.2) can in principle take place in any of the 4 phases. The exact timing usually depends on storage capabilities for liquid refrigerant in the condenser and the need to keep the filling level constant in the evaporator.

In the following, key performance indicators of adsorption chillers are introduced.

Performance indicators

The commonly used indicators to evaluate adsorption chiller performance are the coefficient of performance (COP) and the specific cooling power (SCP) (Sharafian and Bahrami, 2014).

The COP describes the efficiency of the process by relating the desired cooling energy Q_E to the necessary heat energy inputs Q_{IH} and Q_{Des} to run the adsorption chiller cycle:

$$\mathrm{COP} = \frac{Q_E}{Q_{\mathrm{IH}} + Q_{\mathrm{Des}}}. \tag{2.4}$$

As only heats are considered for the calculation of the COP, additional electrical power for control, valves and pumps for secondary circuits are neglected.

The SCP is a measure for the power density. In contrast to COP, SCP additionally considers the factor time and a characteristic reference quantity. The SCP is calculated as the fraction of the desired cooling energy Q_E and the overall cycle time $\Delta t_{\mathrm{cycle}}$ as well as a characteristic mass, usually the adsorbent mass m_{sor}:

$$\mathrm{SCP} = \frac{Q_E}{\Delta t_{\mathrm{cycle}} m_{\mathrm{sor}}}. \tag{2.5}$$

The overall cycle time $\Delta t_{\mathrm{cycle}}$ is the time for a full adsorption/desorption cycle (cf. Figure 2.2).

There is a trade-off between maximising both COP and SCP (Chua et al., 1999). The trade-off leads to Pareto-optimal points, in which the value in one of the dimensions cannot be increased without decreasing the value in another dimension for a given adsorption chiller. The Pareto-optimal points result in a Pareto-frontier which is exemplarily shown in Figure 2.4 and can be used to evaluate and compare the performance of adsorption chillers.

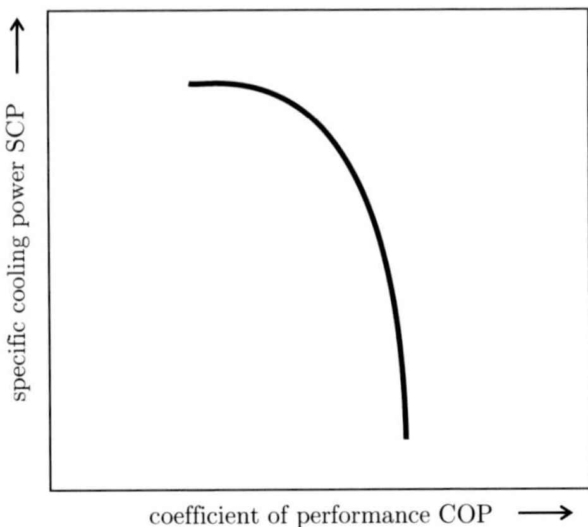

Figure 2.4: Exemplary Pareto-frontier to visualise the trade-off between the 2 main performance indicators for adsorption chillers: SCP vs. COP.

2.1.2 Water-based adsorption chillers: advantages and limitations

One of the main advantages of adsorption chillers is the possibility to use the natural refrigerant water (R-718). Water offers many advantages compared to other refrigerants, in particular the refrigerants used in compression chillers, which are affected by the European F-Gas regulation (European Parliament, 2014).

The natural refrigerant water

- is environmentally absolutely safe: no flammability, low toxicity and GWP = 0 (Heinrich et al., 2015; Foster et al., 2007);
- is not affected by the F-Gas regulation;
- has good availability;
- is cheap and
- has an exceptionally high enthalpy of vaporisation (Wang et al., 2009).

Mechanical compression of the refrigerant water is very challenging and so far – to the best of the author's knowledge – only one water-based compression chiller is commercially available (Efficient Energy GmbH, Feldkirchen, Germany). Simple thermal compression used by adsorption chillers, however, also allows to exploit the outstanding advantages of the natural refrigerant water for cooling applications.

Limitations of the natural refrigerant water

Using water as refrigerant also creates challenges which need to be addressed for successful technological solutions:

Water has a **low saturation pressure** and, therefore, requires vacuum-tight constructions for all components of adsorption chillers to prevent intrusion of inert gases from the surroundings. Furthermore, the triple point of water at 273.16 K (Preston-Thomas, 1990) and 6.116 57 mbar (Guildner et al., 1976) determines the lowest possible temperature at which water in equilibrium is still liquid. Freezing prevents the transport of the refrigerant within the adsorption chiller and results in failure of the chiller. Thus, **operation of adsorption chillers close to or below the triple point of water is infeasible** (Freni et al., 2015; Wang et al., 2009, 2014b). However, many cooling applications require cooling temperatures below the freezing point of water.

Cooling applications at low temperatures, which cannot be fulfilled with water-based adsorption chillers today, include, for example, process cooling in the chemical industry, air desiccation by refrigeration and normal cooling in the food industry, especially dairy farms (Heinrich et al., 2015).

The potential for waste-heat-driven cooling is substantial: according to Heinrich et al. (2015), 13 % of the cooling demand in industrial processes and 8 % in the food industry could be supplied by thermal cooling utilising available waste heat. If waste heat from additional combined heat and power plants is included in the analysis, additional 40 % of the cooling demand in industrial processes and 33 % in the food industry could also be covered by thermal cooling.

Another major challenge for the widespread implementation of adsorption chillers is power density (SCP): power density reflects the cooling power of adsorption chillers, yet is limited by low heat and mass transfers (Li et al., 2015). To improve heat and mass transfer, besides adsorber design, the evaporator design is very important (San and Tsai, 2014; Lanzerath et al., 2015): the maximum pressure during adsorption and therefore the maximum loading (and maximum SCP as well as COP) are affected by the evaporator's performance. Consequently, poor evaporator performance results in poor performance of adsorption chillers (Lanzerath et al., 2016; Lanzerath, 2014; Khan et al., 2007) and calls for better evaporator designs.

More details on the purpose and current state-of-the-art of evaporators in adsorption chillers are given in the next section.

2.1.3 Evaporators in adsorption chillers

The evaporator in an adsorption chiller links the cooling demand (i. e. decreasing the temperature of the heat exchanger fluid) to the adsorption process. This link is realised by transferring heat from the heat exchanger fluid in the secondary circuit to the refrigerant in the evaporator which is vaporised. The generated refrigerant vapour should have a high pressure since higher vapour pressures during adsorption increase the maximum loading and are thus beneficial for the performance of the adsorption cycle (cf. Figure 2.3).

Figure 2.5 shows the integration of the evaporator (right) into the adsorption chiller (left). The reflux from the condenser supplies liquid refrigerant to the evaporator and the evaporated refrigerant leaves the evaporator through a steam connection to the adsorber.

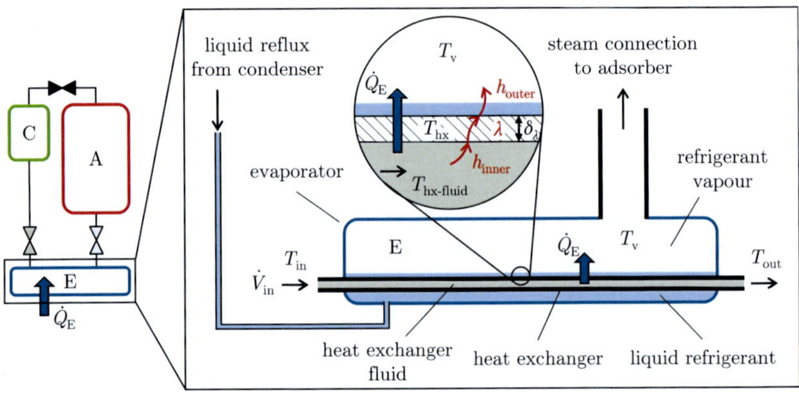

Figure 2.5: Integration of the evaporator in adsorption chillers and enlarged part view of 3-step heat transfer through the evaporator heat exchanger.

The heat exchanger in the evaporator (exemplarily shown as a tube in Figure 2.5) separates the refrigerant from the secondary circuit and the surroundings and ensures vacuum-tightness while still enabling the heat flow $\dot{Q}_E$ necessary for the phase change of the refrigerant.

Often, the overall heat transfer coefficient U defined by

$$U = \frac{\dot{Q}_{\mathrm{E}}}{A \Delta T} \tag{2.6}$$

is used to quantify the ability of a heat exchanger to transfer heat: higher U-values represent a lower thermal resistance for the heat flow $\dot{Q}_{\mathrm{E}}$ and, thus, better heat transfer (Baehr and Stephan, 2011). In Equation (2.6), A is the reference area through which the heat is transferred. The driving force for the heat flow $\dot{Q}_{\mathrm{E}}$ is the temperature difference ΔT, e. g. the temperature difference between the temperatures of the heat exchanger fluid $T_{\mathrm{hx-fluid}}$ and the refrigerant vapour T_{v} (cf. enlarged part view in Figure 2.5).

As depicted in the enlarged part view in Figure 2.5, the heat transfer can be divided into a series of 3 steps (Baehr and Stephan, 2011):

- inner convective heat transfer from the heat exchanger fluid to the heat exchanger wall,
- heat conduction through the heat exchanger wall and
- outer convective heat transfer from the heat exchanger wall into the refrigerant.

For the simple case of a flat plate, the resulting overall heat transfer coefficient U_{plate} for the heat flow $\dot{Q}_{\mathrm{E}}$ through the area A from the heat exchanger fluid to the refrigerant vapour can be calculated by

$$U_{\mathrm{plate}} = \frac{1}{A} \left[\frac{1}{h_{\mathrm{inner}} A} + \frac{\delta_{\lambda}}{\lambda} + \frac{1}{h_{\mathrm{outer}} A} \right]^{-1} \tag{2.7}$$

with plate thickness δ_{λ} and thermal conductivity λ of the plate as well as the 2 convective heat transfer coefficients h_{inner} and h_{outer}. The local convective heat transfer coefficients h_{i} are given by

$$h_{\mathrm{i}} = \frac{\dot{Q}_{\mathrm{i}}}{A_{\mathrm{i}} \Delta T_{\mathrm{i}}} \tag{2.8}$$

with their respective local temperature difference ΔT_{i}.

The magnitude of the 3 heat transfer steps should be balanced, as a poor heat transfer in one of the serial steps can severely limit the overall heat transfer (cf. Equation (2.7)). Often, the thermal conductivity λ of the used materials (e. g. copper) is very high and the wall thickness δ_{λ} of the heat exchanger is comparably thin. Thus, the temperature gradient in the wall in the direction of the heat flow is small and the resistance due to heat conduction can be neglected. Consequently, the 2 remaining convective heat transfers often govern the overall heat transfer coefficient.

The inner convective heat transfer usually depends on the characteristics of the flow of the heat exchanger fluid through the heat exchanger as well as the heat exchanger's inner surface. High inlet volume flow $\dot{V}_{\text{in}}$ (Figure 2.5) as well as turbulence structures can create turbulent flows and increase the inner heat transfer. However, the necessary exergetic effort (e. g. power for pumping the heat exchanger fluid) to provide good inner heat transfer needs to be considered for developing a sound heat exchanger design. According to Witte (2016), the inner convective heat transfer does usually not limit current evaporator designs in adsorption chillers.

The outer heat transfer from the heat exchanger to the refrigerant is crucial for high overall heat transfer coefficients and is, therefore, the main objective of current heat exchanger research, including this thesis. Theoretical fundamentals of the outer heat transfer for evaporation are presented in Section 2.2.

Other design choices also affect evaporator performance: particularly in setups in which the evaporator is also used as condenser (e. g. valveless adsorption chiller designs with only one vacuum chamber) it is also important that the evaporator has a low thermal mass (Lang et al., 1999). A lower thermal mass enables quicker temperature changes and, thus, also shorter start-up times. The thermal mass can be reduced by using both less heat exchanger material and heat exchanger fluid in the evaporator design.

Especially for mobile applications, packaging is also important for adsorption chiller design, including the evaporator. The space required for the heat exchanger as well as the volume of refrigerant liquid and vapour should be as small as possible (Witte, 2016). However, for small cross-sections and low densities, the velocity of the vapour flow can easily be in the range of the speed of sound. Therefore, the velocity should be evaluated considering the flow resistances along the path of the vapour flow. Compactness of the heat exchanger can be considered by choosing the reference area A of the U-value in Equation (2.6) accordingly, e. g. a surface enclosing the heat exchanger to account for its volume (Lanzerath et al., 2016).

Desirable characteristics of evaporators for adsorption chillers can be summarised by describing an "ideal evaporator".

Ideal evaporator

In general, the purpose of an evaporator is to cool down the heat exchanger fluid as far as possible while maintaining a high saturation pressure of the evaporated refrigerant. Thus, an ideal evaporator enables a maximum heat flow $\dot{Q}_{\mathrm{E}}$ with the smallest driving temperature difference ΔT across the heat exchanger wall and the smallest heat transfer area A, resulting in a maximum U-value (cf. Equation (2.6)). Furthermore, the ideal evaporator's construction is as compact as possible, withstands the forces of vacuum conditions while being absolutely vacuum-tight, has minimal hydraulic resistance for the flow of the heat exchanger fluid and minimal thermal mass.

Many of the characteristics of an ideal evaporator contradict each other, e. g. vacuum construction vs. low thermal mass or high inner heat transfer vs. minimal hydraulic resistance, highlighting the challenges in evaporator design. Additionally, in the context of commercial applications, costs of the evaporators become pivotal for design choices and call for simple, inexpensive structures.

Current designs of evaporator heat exchangers

Commercial activities regarding evaporator development usually stay confidential, yet applications for patents involving evaporator heat exchangers have been filed by research-oriented institutions like

- Massachusetts Institute of Technology (McKay et al., 2020),
- Fraunhofer Gesellschaft (Baumeister and Weise, 2017),
- Technische Universität Darmstadt (Schweizer et al., 2011) and
- Bayerisches Zentrum für Angewandte Energieforschung (Costa et al., 2006)

as well as by companies like

- Bayerische Motoren Werke AG (Friedrich, 2017),
- Ford Global Technologies LLC (Levin et al., 2017),
- Vaillant GmbH (Spahn et al., 2016),
- InvenSor GmbH (Braunschweig et al., 2016),
- Fahrenheit GmbH (Mittelbach and Dassler, 2014),
- Mahle Behr GmbH und Co. KG (Burk and Zwittig, 2010) and
- Bosch Thermotechnik GmbH (Faust, 2001).

Although details of heat exchanger designs and function tend to be concealed in patent publications, the existence of patent applications demonstrates strong commercial interest in evaporator development. A selection of 18 patents containing heat exchangers suitable for evaporators of adsorption chillers is given in Table A.1 in Appendix A.

Further information on state-of-the-art heat exchanger designs currently used in evaporators of adsorption chillers is scarce in the literature. Witte (2016) states that the discontinued, commercial adsorption heat pump by Viessmann Werke GmbH & Co. KG uses a falling-film evaporator while the adsorption heat pump by Vaillant Deutschland GmbH & Co. KG employs a pool-boiling setup with spirally arranged, stainless-steel, corrugated pipes as heat exchanger. In line with these references, Schnabel et al. (2008) states that "existing evaporators mainly use pool-boiling or falling-film concepts".

Falling-film evaporators use a pump to create a thin film on the heat exchanger surface. However, pumps require additional power and moving parts inside the vacuum construction for its operation. Therefore, integrating a pump into an adsorption chiller is not favourable, as one of the major advantages of adsorption chillers is to only require thermal energy. Thus, for adsorption chillers, mainly pool-boiling setups are investigated in the literature, and this thesis also focuses on pool boiling.

For pool boiling, many heat exchangers used in adsorption chillers offer structures to increase the available heat transfer area, e. g. tubes with enhanced surfaces, finned tubes, fin-and-tube heat exchangers, plate heat exchangers or plate-fin heat exchangers. The development of evaporators for pool boiling requires a deeper understanding of the underlying evaporation phenomena. Therefore, the relevant basics of evaporation are introduced in the next section.

2.2 Evaporation: heat transfer and phase change

The fundamentals of evaporation presented in this section are based on the textbook by van Carey (2018) and – due to a slight difference between the versions – on the English and German textbooks by Baehr and Stephan (2011, 2013).

Usually, the phase change from liquid to vapour phase is referred to as "vaporisation" or "boiling" and the term "evaporation" is only used for a certain type of phase change: according to Baehr and Stephan (2011) evaporation is "stagnant boiling" where the phase change takes place at the surface and no bubbles occur. Since only this type of boiling is relevant in this thesis, the term "evaporation" is used for the process of phase change from liquid to vapour phase throughout the thesis; yet in some instances, the more general term "boiling" would also be applicable.

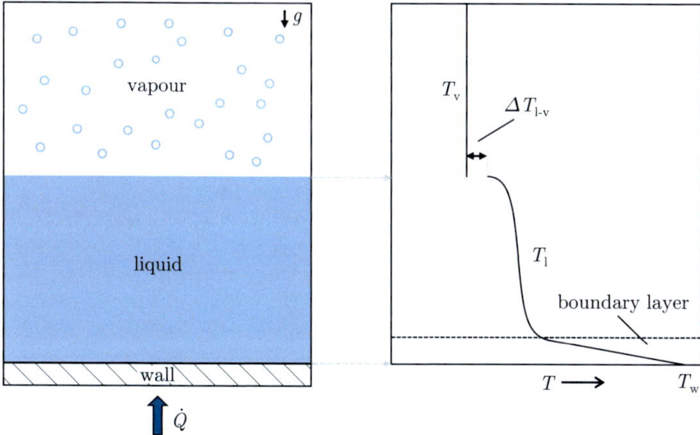

Figure 2.6: Scheme (left) and temperature distribution (right) of evaporation in pool boiling redrawn after Baehr and Stephan (2013). The temperatures including the temperature difference $\Delta T_{l\text{-}v}$ are not to scale, e. g. for water, $\Delta T_{l\text{-}v}$ is around 0.03 K (Prüger, 1940) and it is reasonable to assume that the evaporation process takes place at the saturation temperature of the vapour phase T_v

Liquid–vapour phase change of pure substances at constant pressure is a "virtually isothermal heat transfer" (van Carey, 2018) process. The phase change usually involves high heat transfer coefficients for vaporisation and condensation compared to heat transfer by conduction and hence is attractive for thermal processes, e. g. heat pipes or adsorption chillers.

As stated in the previous section, evaporators in adsorption chillers mainly employ pool boiling. The term "pool boiling" refers to a setting in which the dimensions of the used vessel are large compared to possible bubble sizes and a heated wall introduces a heat flow into the free-flow liquid phase from below (left in Figure 2.6). Although the evaporators in adsorption chillers usually do not fulfil the requirements of this definition in terms of dimensions, the term is commonly used anyway, since the liquid phase is indeed in free flow. During evaporation of a liquid phase in free flow ("stagnant boiling"), the phase change takes place at the surface of the liquid and the enthalpy of vaporisation required for the molecules to change phases is transferred from the heated wall to the surface by natural convection.

The resulting characteristic temperature distribution in pool boiling (cf. Figure 2.6, right) shows a decrease in temperature from the hot wall (T_w) through the liquid phase (T_l) into the vapour phase (T_v). The steep temperature decrease within the boundary layer (in the order of 1 mm) starting from the wall is caused by typically low heat conductivities of liquids. The temperature of the bulk liquid phase (T_l) above the boundary layer is much more homogenous due to developed convective streams. Another steep decrease in temperature is present just below the free surface. Additionally, there is a temperature jump between the temperature at the surface and the saturation temperature of the vapour phase (T_v): the temperature difference $\Delta T_{l\text{-}v}$ (not to scale in Figure 2.6, right) is around 0.03 K for water (Prüger, 1940). Thus, the saturation temperature T_v is almost identical to the evaporation temperature and for technical applications, it is reasonable to assume that the evaporation process takes place at the saturation temperature of the vapour phase.

For evaporation, the heat transfer coefficient is usually very low. However, if the wall temperature (T_w) is increased, bubbles start to form at the heated wall, resulting in higher heat transfer coefficients and a new boiling regime – nucleate boiling – occurs. Different regimes of pool boiling are best described by the Nukiyama curve (Nukiyama, 1966).

2.2.1 Different regimes of pool boiling: Nukiyama curve

The "N-shaped" Nukiyama curve, named after the author Nukiyama (1966), shown in Figure 2.7 is a plot of the heat flux $\dot{q} = \dot{Q}/A$ over the driving temperature difference "wall superheat" $T_w - T_{sat}$ of a pool-boiling setup in which it is assumed that the liquid is at saturation temperature T_{sat}.

As described above, the boiling regime of evaporation ("stagnant boiling", light blue part of the curve in Figure 2.7) is mostly present in evaporators of this work. In this regime, no bubbles occur and natural convection transfers the heat to the surface where evaporation takes place. The gradient of the Nukiyama curve is small. Thus, the heat flux increases only slowly with increasing wall superheat.

At the onset of nucleate boiling (ONB), the wall superheat is sufficient for bubbles to form at the heated surface. Buoyancy takes the bubbles to the surface, thereby inducing intense motion into the liquid and promoting heat transfer. Thus, bubble diameter and departure frequency are important to describe nucleate boiling. Usually, nucleate boiling is the preferred regime of pool boiling in technical applications as it offers high heat transfer coefficients.

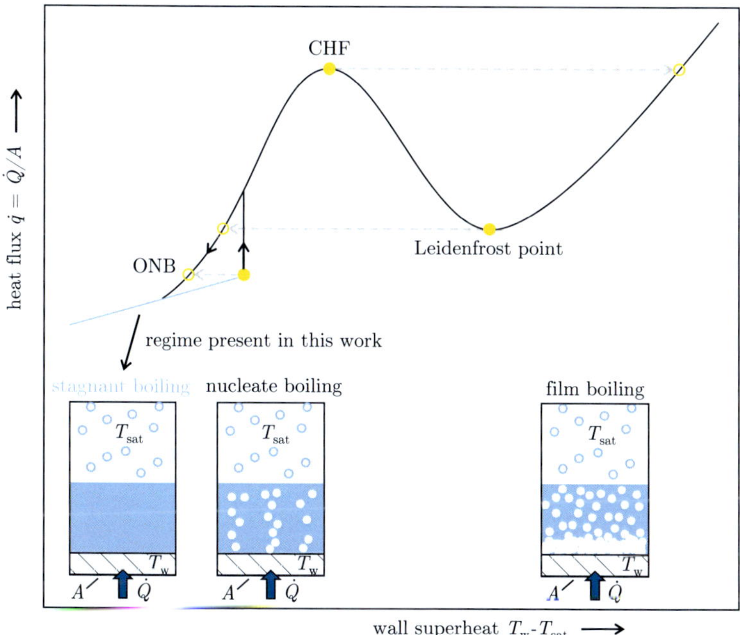

Figure 2.7: Different boiling regimes in pool boiling: Nukiyama curve. Heat flux $\dot{q}$, calculated from heat flow $\dot{Q}$ and area A, is plotted over the driving temperature difference $T_{\text{w}} - T_{\text{sat}}$, calculated from wall temperature T_{w} and saturation temperature T_{sat} of the vapour phase. Also shown are drawings of the boiling regimes present at the wall superheat. The onset of nucleate boiling (ONB) separates the boiling regimes of stagnant and nucleate boiling. The critical heat flux (CHF) is the highest possible heat flux in nucleate boiling. The figure is based on van Carey (2018).

In setups that control the wall temperature T_{w} (cf. $T_{\text{w}} - T_{\text{sat}}$ plotted on horizontal axis in Figure 2.7), pool-boiling experiments follow the solid, black line. Experiments in heat-flux-($\dot{q}$)-controlled setups (heat flux $\dot{q}$ plotted on vertical axis in Figure 2.7), however, follow the dashed, grey line. The hystereses in the process for in- respective decreasing values are indicated by the directions of the arrows; e. g. the ONB starts at a higher wall superheat than the wall superheat at which nucleate boiling ends: at decreasing wall superheats, the last nucleation site stops at a lower wall superheat.

The critical heat flux (CHF) marks the highest possible heat flux $\dot{q}$ in the regime of nucleate boiling.

In technical applications, knowledge of the CHF is important, as for heat-flux-controlled heating surfaces a slight increase in heat flux results in a strong increase in wall superheat, which may destroy the heating surface. Temperature-controlled heating surfaces show a decreased heat flux for increasing wall superheat beyond CHF due to the presence of vapour on parts of the heating surface ("transition boiling").

At the Leidenfrost point, the total heating surface is covered with vapour and the boiling regime is called "film boiling" accordingly. At increasing wall superheat, other forms of heat transfer (e. g. radiation) become more dominant and the heat flux rises again. More details regarding boiling regimes and the involved hystereses can be found in van Carey (2018).

2.2.2 Wetting: interaction of fluids and solids

Heat transfer with phase change is strongly influenced by the way the fluid refrigerant and the solid heat exchanger interact (van Carey, 2018): the interaction depends on intermolecular forces (e. g. van der Waals forces) between the three involved phases liquid, vapour and solid (adhesion) and within the three phases themselves (cohesion). The equilibrium of the involved forces determines the **wetting** behaviour of fluids on the solid and can produce any states from droplets of liquid which sit on the solid's surface to liquids that fully cover a solid's surface with a thin film.

A macroscopic quantity to describe the affinity between liquids and solids and, thus, the wetting behaviour, is the contact angle. The **contact angle** is the angle within the liquid between the liquid–vapour interface and the liquid–solid interface (cf. Figure 2.8, top left). A liquid is considered a wetting liquid, if the contact angle is less than 90° ($\theta < 90°$). If the liquid is water, the solid's surface is then considered hydrophilic. For contact angles $\theta > 90°$, the liquid is referred to as non-wetting; in case of water, the solid's surface is then called hydrophobic.

The contact angle is affected by many factors, including geometry and surface chemistry. Surface chemistry, however, is usually not homogeneous and it is important to note that the surface properties of the solid are only relevant at the location of the **3-phase contact line** (Gao and McCarthy, 2007), the points where all 3 phases (liquid, vapour and solid) meet, e. g. for a droplet sitting on a solid's surface, the contact line is the circle around the wetted area of the solid.

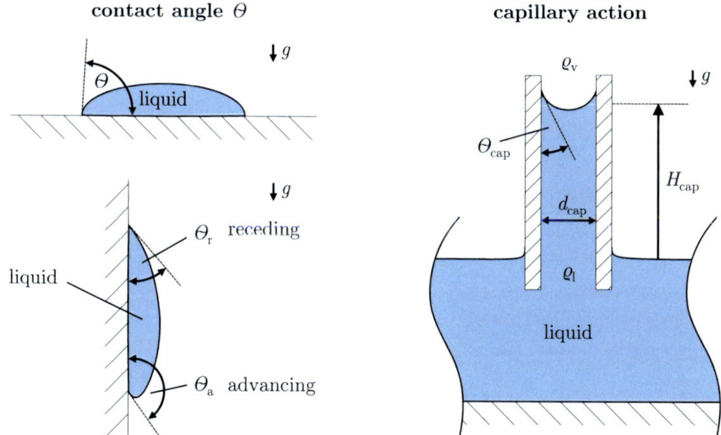

Figure 2.8: Wetting behaviour: interaction between heat exchanger solid phase and refrigerant fluid. Definition of contact angle θ (left, top) and contact angle hysteresis (left, bottom) as well as capillary action (right) with the diameter of the capillary d_{cap}, maximum equilibrium height of the liquid column H_{cap} and the densities of refrigerant vapour ϱ_v and liquid ϱ_l. The direction of gravity g is also shown.

In reality, contact angles are affected by hysteresis, which can be observed when a droplet on a vertical wall is deformed by gravity: the resulting receding (smallest) and advancing (largest) observed contact angle are shown in Figure 2.8 (left, bottom). The observed hysteresis of contact angle can also be caused by local inhomogeneity of the surface (e. g. roughness or impurities). Receding contact angle θ_r and advancing contact angle θ_a are often inaccurately referred to as dynamic contact angles, though they are usually determined quasi-statically (Liu et al., 2019). Nevertheless, contact angles are also affected by dynamics, i. e. the velocity, of the 3-phase contact line: according to Eral et al. (2013), the smallest receding and the largest advancing contact angle are usually determined at small and large capillary numbers Ca, respectively:

$$\mathrm{Ca} = \frac{\eta v}{\sigma_{\text{l-v}}}. \tag{2.9}$$

Here, η is the dynamic viscosity of the liquid, v the velocity of the contact line and $\sigma_{\text{l-v}}$ the interfacial tension between liquid and vapour.

The determination of contact angles proves to be difficult, since many influencing, experimental factors (e. g. temperature, pressure, droplet volume size, how the droplet was placed on the surface) need to be accounted for. Additionally, the existence of receding and advancing contact angles also needs to be considered and even identical setups may not yield the same results (Liu et al., 2019). Furthermore, the contact angle is only defined for macroscopically flat surfaces which excludes many surfaces from quantitative experimental analysis.

Liquid–solid combinations with high or low contact angles (and accordingly liquid–vapour–solid interactions) show interesting phenomena, which can be exploited in technical applications. Low contact angles, for example, enable liquids to rise in structures with small diameters, i. e. capillaries: **capillary action** describes the ability of a liquid to flow against the direction of gravity due to forces between the solid, the liquid and the vapour phase and is schematically shown in Figure 2.8, right. According to van Carey (2018), the maximum equilibrium height of the liquid column H_{cap} due to capillary action for a wetting liquid with contact angle $\theta_{\text{cap}} < 90°$ is given by

$$H_{\text{cap}} = 4\sigma_{\text{l-v}} \frac{\cos\theta_{\text{cap}}}{(\varrho_{\text{l}} - \varrho_{\text{v}}) g d_{\text{cap}}}. \tag{2.10}$$

Here, $\sigma_{\text{l-v}}$ is the interfacial tension between the liquid and the vapour phase, ϱ_{l} and ϱ_{v} are the densities of the liquid and vapour phase, respectively, g denotes gravity and d_{cap} is the diameter of the capillary in which the liquid rises (Figure 2.8, right). Thus, capillary action depends on the characteristics of the solid (diameter of the capillary d_{cap}), the characteristics of the fluid (densities of the liquid ϱ_{l} and vapour phase ϱ_{v} and interfacial tension $\sigma_{\text{l-v}}$) as well as the characteristics of the combination of solid and fluid (contact angle θ_{cap}). For contact angles $\theta_{\text{cap}} > 90°$, the level in the capillary is below the pool level and H_{cap} becomes negative.

Characteristics of the solid are the **surface properties** that affect the contact angle and capillaries. The surface properties can be tuned by coatings. **Coatings** are very versatile, since coatings allow to tailor surface properties for any heat exchanger design. Coatings can be characterised by several properties, including surface roughness, surface extension, coating layer thickness and coating porosity. These properties can be determined by graphical analysis of cross-section polishes of the coatings as exemplarily shown in Table 2.1.

Surface roughness is determined by averaging measured maximum valley-to-peak distances (s_i) of five consecutive sampling lengths (each 100 µm) and denoted R_{z} according to Deutsches Institut für Normung e.V. (1990). R_{z} can also easily be determined using electrical contact (stylus) instruments.

Table 2.1: Visualisation of the surface properties of a coating determined by analysis of a cross-section polish. Sketched cross-section polish shows coating material in grey and substrate in black. Measured quantities are marked blue in the cross section and used to calculate the shown surface properties. The surface roughness is described by R_z calculated according to Deutsches Institut für Normung e.V. (1990).

Property	Unit	Visualisation in cross-section polish	Calculation
surface roughness	µm	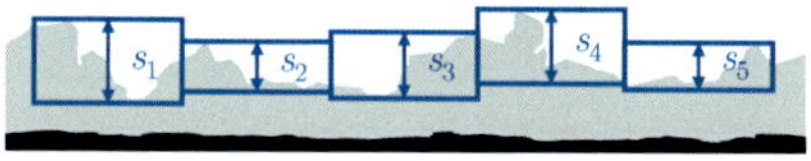 s_1, s_2, s_3, s_4, s_5	$R_z = \frac{1}{5} \sum_{i=1}^{5} s_i$
surface extension	-	 l_1, l_2	$\mathrm{SE} = \frac{l_1}{l_2}$
coating layer thickness	µm	 $\delta_{c,1}$, $\delta_{c,i}$	$\mathrm{LT} = \frac{1}{n} \sum_{i=1}^{n} \delta_{c,i}$
coating layer porosity	%		$\mathrm{CP} = \frac{A_{pore}}{A_{bulk}}$

The **surface extension** is described by the ratio of the length of the surface's contour (l_1) to the shortest possible length (straight line, l_2). Actually, the substrate material onto which the coating is applied (black area in the exemplary cross section in Table 2.1) already exhibits a surface extension by this definition, even without the coating.

The **thickness** of the coating applied to the substrate is arithmetically averaged, as shown in Table 2.1.

The **porosity** of the coating layer can also be estimated by graphical analysis of the cross-section polish: the pore area A_{pore} is detected by image recognition and related to the area of the bulk material A_{bulk}.

The heat exchangers investigated in this thesis use capillary action to transport and spread refrigerant on their surfaces. A similar approach to use capillary action in heat transfer has been used in heat pipes since at least 1944 (Gaugler, 1944). Since some analogies exist, heat pipes are briefly introduced in the next section.

2.2.3 Capillary action applied in heat transfer: heat pipes

Heat pipes are thermal devices which enable heat transfer with high heat flows over long distances at low driving temperature differences. Instead of conducting heat through solid material over the whole distance, heat pipes use the latent heat of a working fluid to transport energy from their evaporator section to their condenser section resulting in extremely high "equivalent thermal conductivity" (Reay et al., 2014) of the entire device. This section on heat pipes is based on textbooks by Faghri (1995) and Reay et al. (2014).

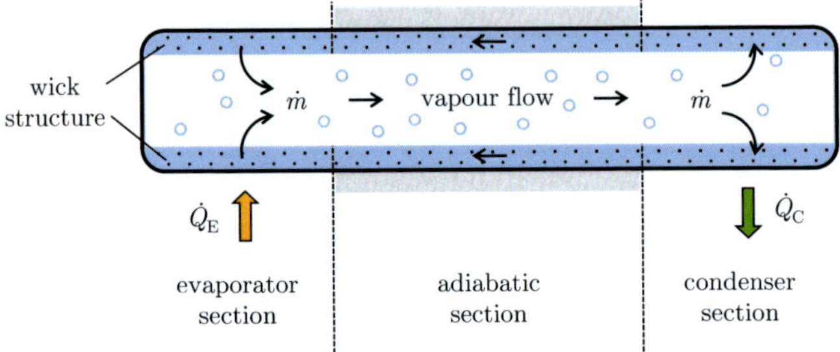

Figure 2.9: Working principle shown in a schematic cross section of a heat pipe. Arrows indicate the direction of working fluid mass flow $\dot{m}$. Heat is absorbed in the evaporator section at a higher temperature (heat flow $\dot{Q}_E$) than the temperature at which the heat is released in the condenser section (heat flow $\dot{Q}_C$). The figure is based on Reay et al. (2014).

The working principle of heat pipes is shown in a schematic cross section of a heat pipe in Figure 2.9. The working fluid evaporates in the evaporator section and the generated vapour mass flow $\dot{m}$ flows towards the colder condenser section. Here, the working fluid condenses and the liquid is transported back to the evaporator through the adiabatic section. A wick structure employs capillary action to transport the liquid working fluid independently of gravity. Thus, heat pipes can be operated in all orientations. Heat pipes, which do not rely on capillary action but transport the liquid working fluid in an inclined setup back to the evaporator by gravity, are called thermosyphons.

The combination of working fluid and wick material needs to be compatible (e. g. wicking, long-term corrosion). The final choice for a compatible combination depends on the temperature at which the heat is transferred: high-temperature applications (up to 2300 °C) may require liquid metals, e. g. lithium, as working fluid together with an expensive, lithium-resistant alloy, while low-temperature applications above 0 °C often use the "ideal working fluid" (Reay et al., 2014) water in combination with copper (Reay et al., 2014).

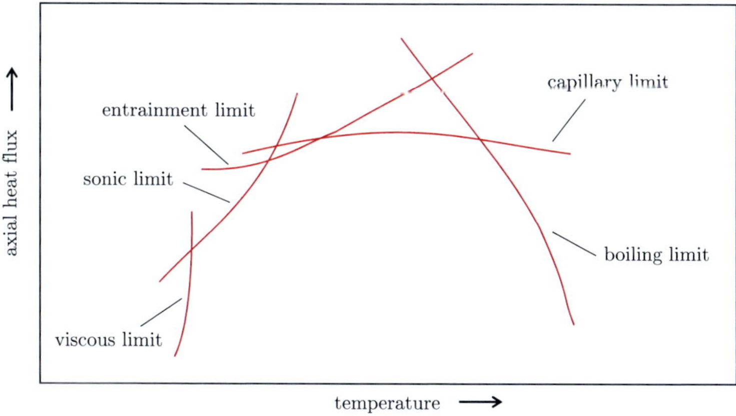

Figure 2.10. Limitations of heat pipes redrawn after Reay et al. (2014).

Often, the evaporator limits the heat pipe's performance (Weibel et al., 2010). Other known limitations of heat pipe performance were first reported by Cotter (1965) and depend on the temperature and axial heat flux through the heat pipe (Figure 2.10) (Reay et al., 2014).

Viscous limit (vapour pressure limit)

At low temperatures (e. g. at start-up), viscous forces in the working fluid dominate the process as the driving pressure difference between evaporator and condenser is low.

Sonic limit

The velocity of the vapour may reach the speed of sound due to low vapour densities, e. g. in liquid metal heat pipes. Hence, a shock wave can be generated and block the vapour flow (Bertossi et al., 2012).

Entrainment limit

At high vapour velocities, vapour can drag liquid along, hindering the liquid to flow back to the evaporator.

Capillary limit

The pressure difference available due to capillary action in the wick needs to overcome the pressure resistances present in the heat pipe: the resistance for the liquid and vapour flow as well as the gravitational head (depending on the orientation of the heat pipe). If capillary action is not sufficient to overcome the resistances, the evaporator section of the heat pump will dry out. Usually, dry-out happens at high mass flows due to high axial heat flows.

Boiling limit

At high radial heat flows in the evaporator section, liquid can boil inside the wick, causing vapour blankets in the evaporator section which prevent wetting of the evaporator and cause hot spots. Finally, the boiling may result in dry-out of the evaporator.

The conditions for evaporation in water-based heat pipes and water-based adsorption chillers are quite similar. In both thermal devices, pure water is evaporated in a closed system. In water-based adsorption chillers, water is evaporated at low temperatures and, thus, at low saturation pressures. At low saturation pressures, efficient evaporation of water with high heat transfer coefficients is challenging. This challenge is explained in detail in the next section.

2.2.4 Challenge: sub-atmospheric evaporation of water

Sub-atmospheric evaporation of water refers to evaporation at pressures between the triple point and ambient pressure. Besides all advantages of the refrigerant water (cf. Section 2.1.2), efficient, sub-atmospheric evaporation of water with high heat transfer coefficients is challenging. In a pool-boiling setup, the heat transfer for evaporation (stagnant boiling) with submerged heat exchanger is limited due to the low heat conductivity and convective heat transfer of water. The favourable nucleate boiling would enable high heat transfer coefficients, however, for sub-atmospheric boiling of water only at the cost of high superheats. One reason for the requirement of high superheats is explained in the following.

Although the process of sub-atmospheric nucleate boiling of water is far more complex (for more details the reader is referred to Witte (2016)), for bubbles to be stable and emerge, the liquid needs to be superheated at least to some degrees, i. e. its temperature needs to be above the saturation temperature (van Carey, 2018). The saturation temperature is a function of the pressure at the submerged, heated wall, where the bubbles emerge. There, the local pressure is affected by the hydrostatic pressure of the water column above (Figure 2.11): a water column of $H_\mathrm{l} = 10\,\mathrm{cm}$, for

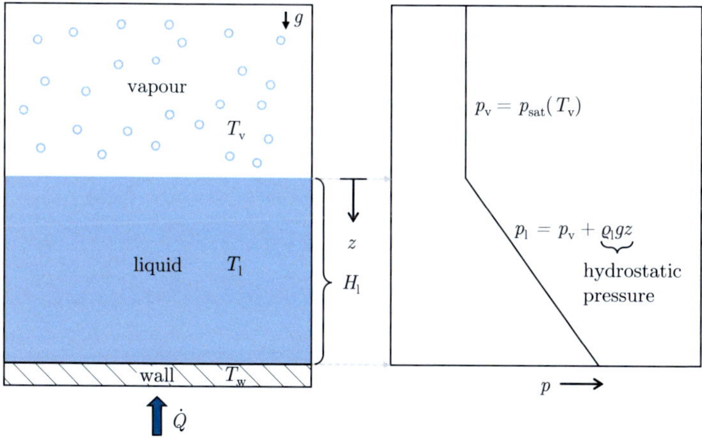

Figure 2.11: Sub-atmospheric evaporation: impact of the static head on the local pressure in the liquid phase.

example, corresponds to a static head of $\Delta p = 9.81$ mbar. The saturation pressure of water decreases strongly with decreasing temperature. Thus, the static head has the strongest effect at low saturation temperatures $T_{sat}(p)$, as can be seen in Figure 2.12. In Figure 2.12, the difference of the saturation temperature at the surface $T_{sat}(p)$ and the local saturation temperature $T_{sat}(p + \varrho g H_l)$ below the water column is plotted. At temperatures close to 100 °C, the difference is almost negligible, while at lower temperatures close to the triple point, the difference increases strongly – even for small water column heights.

As a result, experiments show that nucleate boiling of water at low, sub-atmospheric pressures occurs mainly at high superheats (McGillis et al., 1991) and even metal fibre structures to promote nucleation reduce the required superheats to only just below 7 K (Witte, 2016). Additionally, the induced bubbles are comparably large (diameter up to 15 cm at 8.5 mbar) and boiling is characterised by strong wall temperature fluctuations (Giraud et al., 2015).

However, transferring heat with high superheats is thermodynamically always inefficient. Furthermore, high superheats lead to lower evaporation temperatures that reduce the efficiency of adsorption cooling cycles.

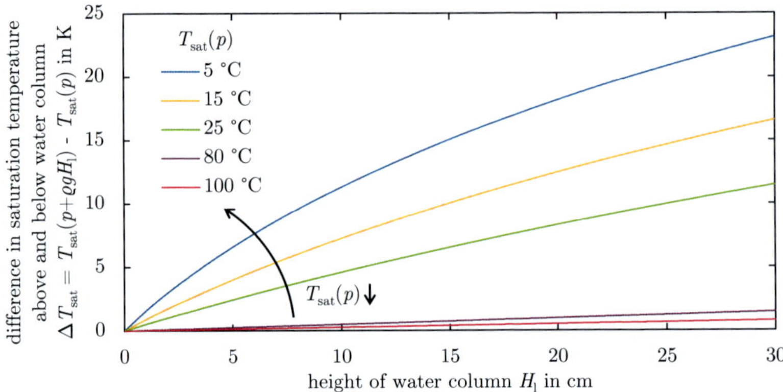

Figure 2.12: Difference in saturation temperature above and below a water column ΔT_{sat} as a function of column height H_l for different saturation temperatures $T_{sat}(p)$.

In order to evaporate water at low temperatures with small superheats, the water column above the heat exchanger can be reduced until only a thin film is left: thin-film evaporation of water combines high heat transfer coefficients with small superheats (Köroğlu et al., 2013).

Maintaining a thin, evaporating film on the heat exchanger is a challenge. The classical, state-of-the-art solution is a falling-film evaporator, in which the refrigerant is sprinkled on the heat exchanger with a pump. However, active pumping and distribution of a saturated liquid at low pressures (I) suffers from non-uniform liquid distribution (Wang et al., 2011) (II) increases the complexity of the system (Xia et al., 2008) and (III) relies on additional external exergy for operation. Furthermore, own experiments involving pumping of saturated water at low pressure indicated that preventing cavitation can be challenging.

The mentioned drawbacks can be avoided by substituting the mechanical pump by capillary action to distribute the refrigerant on the heat exchanger. Thus, capillary-assisted thin-film evaporators are well suited for sub-atmospheric evaporation of water in adsorption chillers. A summary of previous work on capillary-assisted, thin-film evaporators is given in the next section.

2.2.5 Capillary-assisted evaporators for adsorption chillers

Capillary-assisted evaporators exploit capillary action to maintain a thin film during evaporation on the heat exchanger. The most common application of capillary-assisted evaporators is in heat pipes (cf. Section 2.2.3).

However, capillary-assisted evaporators are also investigated and applied in water-based ab-/adsorption chillers and heat pumps (Chen et al., 2010). To the best of the author's knowledge, the first experimental investigation of a water-based capillary-assisted evaporator beyond heat pipes was in a falling-film heat exchanger for an absorption chiller (Sabir and Bwalya, 2002). As mentioned above, capillary action is used in falling-film evaporators to improve the liquid distribution and to increase the heat transfer coefficient (Kim and Kang, 2003; Sabir and ElHag, 2007; Sabir et al., 2008; Lee et al., 2012; Köroğlu et al., 2013; Dang et al., 2017), yet the creation of the falling film still requires external exergy for pumping of the refrigerant.

The focus of this thesis lies on evaporators for adsorption chillers, which solely rely on capillary action for distributing the refrigerant on the heat exchanger and do not require any additional external exergy for pumping. Nevertheless, evaporators which solely work with capillary action could most likely also improve the refrigerant distribution and, thus, heat transfer in falling-film heat exchangers.

To create the necessary capillary action for maintaining a thin film on the heat exchanger, different structures have been suggested in the literature:

(1) **Macroscopic structures**, e. g. finned tubes (Xia et al., 2008; Lanzerath et al., 2016; Thimmaiah et al., 2016), fin-and-tube (Volmer et al., 2014, 2015, 2017), plate-fin and tube-wire (Warlo et al., 2017; Schnabel et al., 2018), and metallic short fibre structures (Witte, 2016),

(2) **Microscopic structures**, e. g. coatings (Schnabel and Witte, 2009; Schnabel et al., 2010, 2011; Lanzerath et al., 2016; Thimmaiah et al., 2017),

(3) **Combination of macroscopic and microscopic structures**, e. g. coated macroscopic structures (Schnabel and Witte, 2009; Lanzerath et al., 2016).

In addition to different structures, the following characteristics of capillary-assisted thin-film evaporation of water have been investigated in the literature: (I) evaporation temperature, (II) driving force and (III) filling level (Figure 2.13).

(I) The **evaporation temperature** (cf. Figure 2.6) can be estimated by the vapour temperature T_v which corresponds to the saturation temperature T_{sat} and the pressure p. Experimental studies by Xia et al. (2008), Lanzerath et al. (2016) and Thimmaiah et al. (2017) agree that the heat transfer increases with

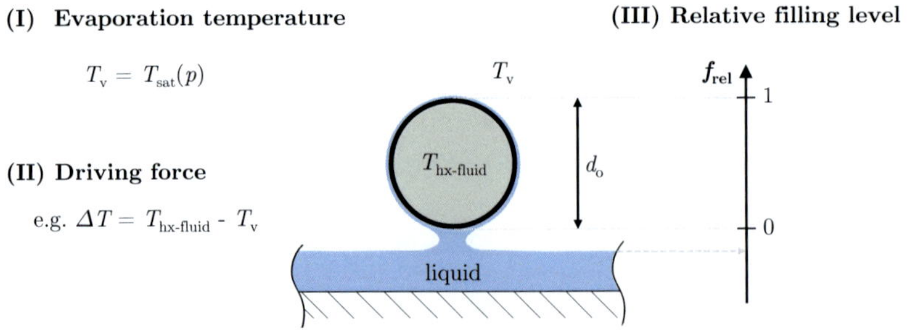

Figure 2.13: Investigated characteristics of capillary-assisted, thin-film evaporation in literature: (I) evaporation temperature, (II) driving force and (III) relative filling level f_{rel} as a function of vertical heat exchanger dimension, here outer tube diameter d_{o}. Exemplarily, a cross section of a structured tube is shown. During evaporation, capillary action continues to wet the tube with liquid fed through the meniscus formed between the tube and the liquid pool.

rising evaporator temperature; it shall be noted that in some studies the effect of evaporation temperature is analysed by variation of the temperature of the heat exchanger fluid $T_{\mathrm{hx-fluid}}$ instead of the temperature of the vapour T_{v} (cf. Figure 2.13). An increase in evaporation temperature improves the physical properties of the refrigerant, which then most likely cause the observed improvement in heat transfer.

(II) The **driving force**, e. g. the temperature difference ΔT, is the force that causes evaporation. The impact of the driving force on heat transfer coefficients has also been examined: for evaporation (stagnant boiling), as introduced in Section 2.2.1, the heat transfer coefficient increases with increasing driving force (wall superheat). However, for thin-film evaporation, reported results differ: for increasing driving force, Xia et al. (2008) detected lower heat transfer coefficients, while experiments by Lanzerath et al. (2016) did not show any impact of the driving force. Pretests in the study of Volmer et al. (2017) also revealed no significant impact of the driving force. Thus, the impact of the driving force does not seem to be fully understood and requires further experimental investigation.

(III) The **filling level**, is the degree to which the heat exchanger is submerged in the refrigerant. The relative filling level f_{rel} is defined relative to the dimension of the heat exchanger, e. g. the outer diameter of the tube d_{o} (cf. Figure 2.13). Submerged parts of the heat exchangers cannot participate in thin-film evaporation. Thus, in general, the potential for thin-film evaporation decreases with higher filling levels. However, the findings regarding the impact of the filling level are not consistent in literature: Xia et al. (2008) and Lanzerath et al. (2016) report increasing heat transfer with lower filling levels, although their experimental procedures for determining the heat transfer at different filling levels differ: Xia et al. (2008) measured at distinct, constant filling levels, while Lanzerath et al. (2016) evaluated all possible filling levels with slowly decreasing filling level in one experiment. In contrast, Thimmaiah et al. (2017) did not find the maximum heat transfer coefficient at the lowest filling level, but at a relative filling level of about 0.8. Volmer et al. (2017) stated that the combination of filling level and heat exchanger geometry affected the heat transfer. These different findings on the impact of the filling level, first of all, confirm the filling level's general importance but also call for further experimental investigation to better understand the impact of the filling level.

Thus, the literature on capillary-assisted thin-film evaporation of water for adsorption chillers reveals inconsistencies for some aspects. Furthermore, comparisons of many published studies prove to be difficult due to experimental differences in the 3 characteristics (I) evaporation temperature, (II) driving force and (III) filling level. Also, other important impacts, like the presence of non-condensable gases in the system (Fischer, 2015), may not have been considered in all studies but affect the outcome of experiments and, thus, question the validity of results. To show validity of results, experiments should be reproducible and, if possible, compared to the results of similar experiments.

In all mentioned evaporators, capillary action is used to keep the heat exchangers wetted with refrigerant during evaporation. However, in addition to the fact that thin films are generally favourable for evaporation of water in adsorption chillers (cf. Section 2.2.4), a mechanistic understanding is also important to improve evaporation in capillary-assisted evaporators. According to Plawsky et al. (2014), "evaporation from thin film is nearly always in the context of an evaporation meniscus". Evaporation from menisci in capillaries has been investigated in the context of heat pipes for a long time: theoretical and experimental investigations suggest that most evaporation takes place in a rather small region close to the 3-phase contact line called **extended meniscus**. Thus, for a general understanding, details on the evaporation mechanisms are given in the next section.

2.2.6 Evaporation from an extended meniscus

Evaporation can occur from extended menisci in both evaporation (stagnant boiling) and nucleate boiling (Figure 2.14). In nucleate boiling, one of the heat transfer mechanisms that is active during bubble formation is thin-film evaporation (Figure 2.14 (a)) (Zou et al., 2016). Relevant research for thin-film evaporation is often in the context of nucleate boiling, since nucleate boiling is widely used in technical applications and of high importance for many industries.

Similar conditions can also be present in evaporation from a capillary as shown in Figure 2.14, (b): the extended meniscus in (b) is very similar to the extended meniscus in (a). Only the orientation and, therefore, the impact of gravity differ in the shown setting.

The 3-phase contact line consists of all points where the three involved phases meet: heat exchanger solid, refrigerant liquid and vapour phase. Thus, the 3-phase contact line is usually located close to the measurement point of the contact angle θ (Figure 2.14, (b)). In the literature, the extended meniscus has been divided into 3 regions (Figures 2.14 and 2.15) (Potash and Wayner, 1972):

I Macroscopic meniscus

II Evaporating thin film

III Non-evaporating adsorbed film

The exact definitions of the transitions from region to region may differ in literature (Crößmann, 2016), yet are not important for the scope of this thesis. In general, the macroscopic meniscus region (I) represents the bulk of the liquid. The curvature of the liquid's interface is almost constant and defines the macroscopically detectable contact angle θ. The evaporating thin-film region (II) is comparably small and characterised by locally high heat fluxes. The remaining region, the non-evaporating adsorbed film (III), can macroscopically not necessarily be identified as being covered by refrigerant (Fischer, 2015): just a very thin layer of refrigerant is adsorbed to the surface and does not evaporate. Thus, the apparent macroscopic 3-phase contact line is located between regions II and III.

According to Plawsky et al. (2014), the increase in heat flux in region (II) close to the contact line in Figure 2.15 can be explained by analysing the involved heat transfer resistances: the resistance due to conductance and the resistance at the liquid–vapour interface result in the overall thermal resistance given in Figure 2.15.

While conductance through the heated wall should be similar in all regions, subsequent conductance through the liquid phase mostly depends on the thickness of

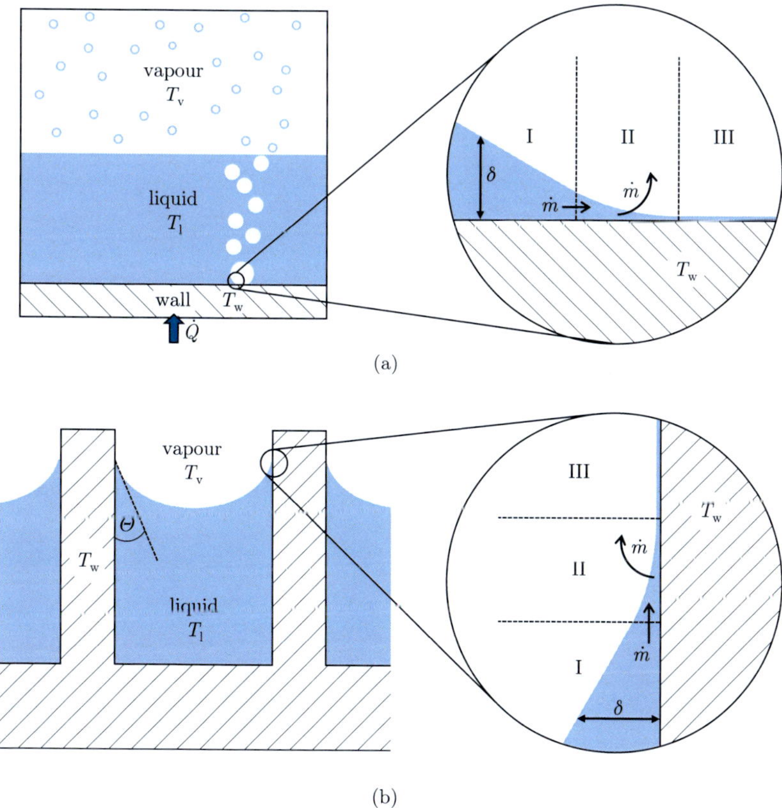

Figure 2.14: Thin-film evaporation in nucleate boiling (a) and evaporation (stagnant boiling) from a capillary (b). The magnifications show the definition of 3 regions in an extended meniscus including the vicinity of the 3-phase contact line. Detailed analysis of the 3 regions is presented in Figure 2.15. The figure is based on Xia et al. (2008), Crößmann (2016) and Potash and Wayner (1972).

the liquid film δ (cf. δ/λ from Equation (2.7) for heat conductance through the heat exchanger). Thus, as the thickness δ decreases from region (I) to (III), resistance due to conductance through the liquid is mainly dominant in region (I).

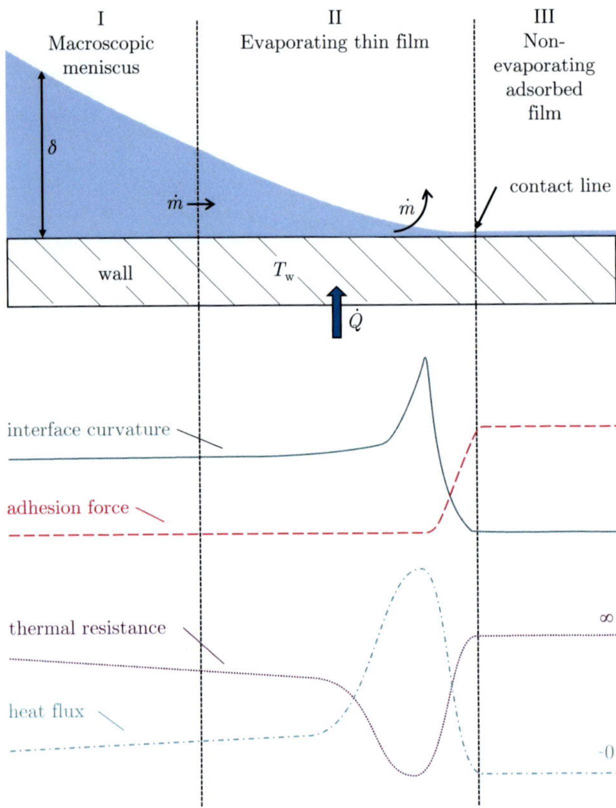

Figure 2.15: Definition of regions in the extended meniscus (cf. Figure 2.14) with liquid thickness δ, mass flow $\dot{m}$, wall temperature T_w and heat flow $\dot{Q}$ for evaporation. Below are spatially-resolved qualitative profiles of the interface curvature between liquid and vapour phase above, the adhesion force exerted from the molecules of the wall onto the molecules of the liquid and the overall thermal resistance from the wall into the vapour for the local heat flux. Figure is redrawn after Plawsky et al. (2014).

The heat transfer resistance at the liquid–vapour interface can be dominating for liquids with a very high enthalpy of vaporisation and is affected by many factors, e. g. the vapour–liquid interface curvature and intermolecular forces (Plawsky et al., 2014).

The intermolecular forces (e. g. van-der-Waals forces) between the liquid and the solid wall result in adhesion of the liquid on the wall and are summarised in the concept of "disjoining pressure". The disjoining pressure p_{dis} is only relevant very close to the wall, since its magnitude is inversely proportional to the third power of the distance δ: ($p_{\text{dis}} \sim \delta^{-3}$) (Crößmann, 2016). Hence, the adhesion force due to the disjoining pressure is mainly dominant in the adsorbed film in region (III) where the thickness δ of the liquid film is very small. Although the adsorbed film has the same temperature as the heated wall, the adhesion forces prevent phase change in region (III) (heat flux $\rightarrow \sim 0$ and thermal resistance $\rightarrow \infty$).

In region (II), however, phase change is possible due to increasing thickness δ and strong changes of the curvature of the vapour–liquid interface. The interface curvature κ and interfacial tension σ affect the capillary pressure ($p_{\text{cap}} = \sigma\kappa$) and, therefore, also the pressure difference between the liquid and vapour phase. For the concave surface of the liquid shown in Figure 2.15, the pressure in the liquid phase is reduced, and, thus, strong evaporation takes place at the maximum interface curvature (Baehr and Stephan, 2011). Furthermore, the combination of lower pressure in the liquid phase and evaporation in region (II) leads to a horizontal mass flow of liquid towards the contact line, thereby intensifying the convective heat transfer from the heated wall into the liquid. Hydrodynamic resistances of this liquid mass flow can also limit evaporation (Plawsky et al., 2014). Thus, according to Plawsky et al. (2014), maximising overall heat transfer needs to include both, enlarging the favourable evaporating-thin-film region (II) and ensuring sufficient transport of liquid into that region.

The evaporating-thin-film region (II) is still under investigation. For the case of bubble formation in nucleate boiling, there has been a scientific debate about the contribution of different heat transfer mechanisms that are relevant for bubble growth (Kim, 2009). Besides the described evaporation in the vicinity of the 3-phase contact line ("contact line evaporation"), other evaporation mechanisms can also be present, e. g. receding liquid underneath a growing bubble can deposit a thin film on the heated wall (microlayer), whose evaporation also contributes to bubble growth (Kim, 2009). During "microlayer evaporation", a layer significantly thicker than in the evaporating-thin-film region (II) in Figure 2.15 is present between regions (II) and (III).

A transition from contact line-dominant evaporation to microlayer-dominant evaporation was detected by Schweikert et al. (2019) for the perfectly wetting fluid FC-72. The experiments revealed that the transition between evaporation regimes occurs in case a "critical velocity is exceeded for a given combination of wall superheat and heating power" (Schweikert et al., 2019). Fischer (2015) found that contact line-dominant heat transfer increases for higher velocities of an advancing contact line while the heat transfer is not affected by the velocity of a receding contact line. Apparently, the velocity of the contact line is important for heat transfer. Plawsky et al. (2014) concluded in a review that **transient operation** might be key to explore even higher instantaneous and time-averaged heat transfer capabilities for thin-film evaporation.

All presented findings show that heat transfer mechanisms for thin-film evaporation processes are affected by many factors, including contact angle (curvature of the vapour–liquid interface), evaporation temperature, driving force, heating power and the velocity of the contact line.

2.3 Summary and contribution of this thesis: debottlenecking the evaporator in water-based adsorption chillers

Besides utilising low-grade heat to provide environmentally benign cooling, one of the main advantages of adsorption chillers is the ability to use the natural refrigerant water. Water offers outstanding properties as refrigerant for use in adsorption chillers, yet two main bottlenecks have been identified for the refrigerant water:

(I) **Efficient sub-atmospheric evaporation of water** with high heat transfer coefficients is challenging due to its low saturation pressure and the impact of the static head.
In general, to improve evaporator performance and power density, the evaporator's U-value needs to be increased (cf. Section 2.1.3). Despite major advantages of the natural refrigerant water, **efficient evaporation with high U-values** is challenging (cf. Section 2.2.4). Capillary-assisted evaporators address the challenges linked to sub-atmospheric evaporation of water. Yet, many effects in capillary-assisted evaporators for adsorption chillers are unclear, including the **impacts of the driving force**, the **filling level** and **transient operation**. Furthermore, analysis of experimental **reproducibility of capillary-assisted evaporation experiments** is missing in literature (cf. Section 2.2.5).
Capillary-assisted evaporators rely on capillary action. Capillary action is governed by the combination of the fluid, the heat exchanger's design and surface properties. Surface properties can be altered by **coatings**. (cf. Section 2.2.2). Coatings can tailor surface properties for many heat exchanger designs. Correlations between surface properties of the coating and resulting U-values for capillary-assisted thin-film evaporation for adsorption chillers are important for

heat exchanger development but also lacking in the literature.

(II) **Freezing of the refrigerant water at low temperatures** limits its application and, thus, excludes many cooling services: the lowest possible temperature for the refrigerant water is restricted by its triple point because freezing prevents operation of adsorption chiller (cf. Section 2.1.2).
Since many processes require cooling at lower temperatures, **expanding the operational range of water-based adsorption chillers** would allow for environmentally benign cooling in further processes. Additionally, increasing the evaporator's U-value and lowering the superheat required for the heat transfer in the evaporator also enables lower temperatures for cooling applications.

This thesis addresses both identified bottlenecks:

(I) **Efficient sub-atmospheric evaporation of water** is experimentally demonstrated by investigating capillary-assisted thin-film evaporation of water on coated tubes. Impacts of driving force, temperature and filling level are experimentally analysed and the properties of the investigated coatings correlated with determined U-values.

(II) **Freezing of the refrigerant water at low temperatures** in the evaporator of water-based adsorption chillers is overcome by adding the anti-freezing agent ethylene glycol into the evaporator. The ethylene glycol–water mixture in the evaporator stays liquid below 0 °C and, thus, enables water-based adsorption cooling below the freezing point of water.

Both solutions, capillary-assisted thin-film evaporation and addition of an anti-freezing agent, mainly affect the evaporator of the adsorption chiller. Therefore, this thesis focuses on the evaporator of adsorption chillers. The contributions of this thesis are presented in Chapters 3–6. The contents of the chapters are summarised and visualised in the context of an adsorption chiller and its evaporator in Figure 2.16.

Two experimental setups, including measurement equipment and data reduction, have been designed, implemented and used to conduct experiments. Both experimental setups are introduced and described in detail in **Chapter 3**. Evaporation experiments regarding capillary-assisted thin-film evaporation were conducted in a specialised setup that allowed to adjust the inlet temperature, driving force and filling level for evaporation experiments on tubes. Experiments to expand the operational range of water-based adsorption chillers below 0 °C were performed in a one-bed adsorption chiller setup to demonstrate feasibility. Chapter 3 concludes with an introduction to the analysis of uncertainty in measurement and a description of the calibration of the temperature sensors to ensure accuracy for the conducted experiments.

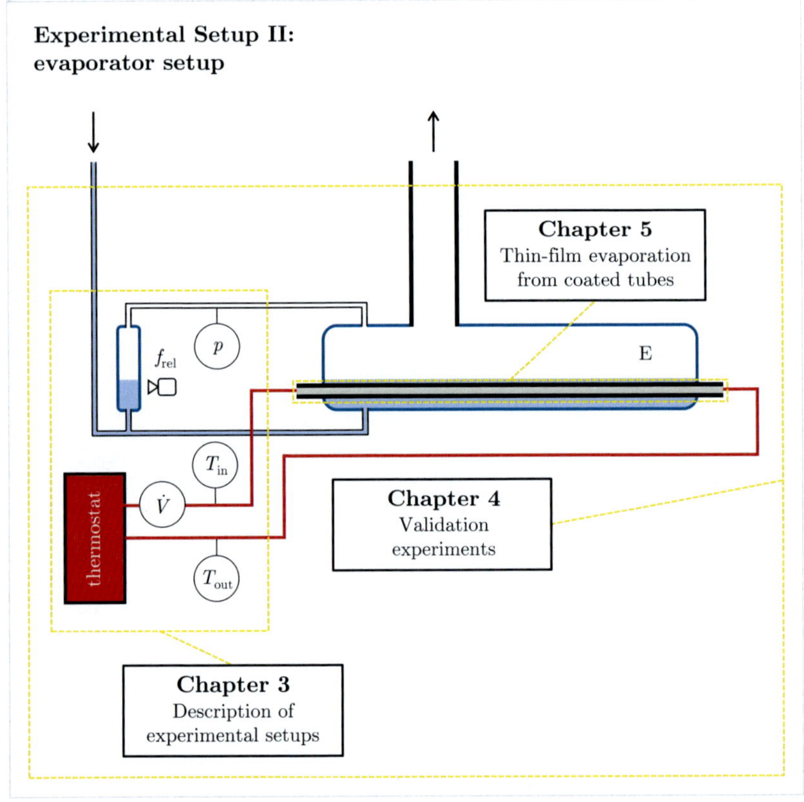
Experimental Setup I:
adsorption chiller setup
C
A
E
Chapter 6
Ethylene glycol as anti-freezing agent
Experimental Setup II:
evaporator setup
Chapter 5
Thin-film evaporation from coated tubes
p
f_{rel}
E
T_{in}
$\dot{V}$
thermostat
T_{out}
Chapter 4
Validation experiments
Chapter 3
Description of experimental setups

Figure 2.16: (figure on previous page) Contribution of this thesis: contents of chapter 3 to 6 exemplarily shown in the context of an adsorption chiller (above) and the evaporator (enlarged view below). The sensors of the evaporation experiments are exemplarily highlighted to represent the description of both experimental setups in Chapter 3. The validation experiments in Chapter 4 are represented by the highlighted evaporator. Thin-film evaporation experiments on the highlighted tube in the evaporator are presented in Chapter 5. The experimental setup with the adsorption chiller represents Chapter 6.

In **Chapter 4**, the experimental setup and procedure for the evaporation experiments is validated: identical experiments with capillary-assisted evaporation on finned tubes were conducted in two separate experimental setups located in two labs – the experimental setup at RWTH Aachen University used in this thesis and an experimental setup at Fraunhofer ISE in Freiburg, Germany. The results are compared and impacts regarding experimental setups and procedures are deduced. Furthermore, impacts of the experimental procedure regarding among others (I) dynamics (constant vs. decreasing filling levels), (II) non-condensable gases and (III) surface properties are analysed and discussed. Thereby, a general guideline to reproducibly conduct sub-atmospheric, thin-film evaporation experiments is derived.

The validated experimental setup for evaporation experiments is used in **Chapter 5** to study capillary-assisted, thin-film evaporation from coated tubes with high U-values. Evaporation performance of 7 coatings and a plain tube is investigated for varying evaporator inlet temperature, driving force and filling level. The investigated coatings differ in porosity, surface extension, roughness and coating thickness. Measured surface properties are correlated to the experimentally determined evaporation performance to provide a basis for tailoring future coatings for capillary-assisted thin-film evaporation. Impacts of evaporator inlet temperature, driving force and filling level on thin-film evaporation are analysed and compared to results from the literature, thereby helping to resolve inconclusive findings in the literature.

In **Chapter 6**, the operational range of a water-based adsorption chiller is expanded below 0 °C by adding ethylene glycol as anti-freezing agent to the evaporator. Experiments are conducted to examine the feasibility of water-based adsorption chiller below 0 °C. The performance is evaluated at different ethylene glycol mass fractions and different cooling temperatures including non-freezing conditions. Thus, comparisons to pure water as refrigerant become possible and allow to identify challenges for the successful implementation of the process.

Chapter 3

Experiments and measurement uncertainty

This chapter contains detailed descriptions of the 2 experimental setups used for the studies reported in this thesis.

The first experimental setup allows to operate and characterise adsorption chillers. A former version of the setup had been used by Lanzerath (2014) to experimentally validate dynamic models of adsorption chillers (Lanzerath et al., 2015). For the work on this thesis, the experimental setup was revised and upgraded to study anti-freezing agents. The experimental setup for adsorption experiments is described in Section 3.1.

The second experimental setup was used to characterise the heat transfer during capillary-assisted evaporation at conditions present in adsorption chillers. An iterative process of design, construction and operation led to the final version of the experimental setup for evaporation experiments. The setup is introduced in Section 3.2.

Finally, analysis of measurement uncertainty that is relevant for both experimental setups is presented in Section 3.3.

3.1 Adsorption experiments

Adsorption experiments require an experimental setup, which provides and measures enthalpy flows that are necessary for the operation of the adsorption chiller. The investigated adsorption chiller is presented in Section 3.1.1. Details of the experimental setup are given in Section 3.1.2. The last Section 3.1.3 introduces the employed data reduction for the data analysis.

3.1.1 Adsorption chiller

The investigated one-bed adsorption chiller was composed of the components adsorber, evaporator and condenser, which were interconnected by valves. All components were vacuum-tight and insulated to prevent unwanted heat flows to and from the surroundings. Heat flows required for the operation of the chiller according to the cycle explained in Section 2.1.1 were transferred by heat exchangers in each component. Specifics of each component are described in the following.

Adsorber

The adsorber consisted of a vacuum-proof barrel, in which a fin-coil heat exchanger was placed. The heat exchanger contained about 5 kg adsorbent (zeolite NaY) between the fins. (Figure 3.1). A vapour-permeable sheeting kept the adsorbent between the fins. Geometric specifications of the adsorber are given in Table B.1 in Appendix B.

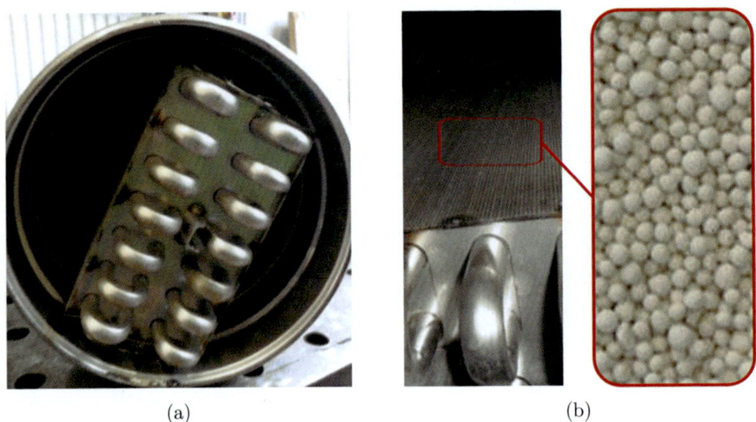

(a) (b)

Figure 3.1: Photos of the adsorber used in the adsorption chiller: bottom view on the heat exchanger in open barrel (a) and close-up of fin-coil heat exchanger covered with vapour-permeable, stainless-steel sheeting and granular zeolite NaY (b).

The adsorber was not specifically designed for the experiment and did not allow for high SCP, but was available and chosen due to the used materials: all parts were made from stainless steel and were thus compatible with many possible water additives, including ethylene glycol. The adsorber was equipped with 2 DN40 ISO-KF flanges at the bottom and the top for steam connection to the evaporator and condenser. Additionally, 2 DN16 ISO-KF flanges at the top of the adsorber allowed to connect pressure and temperature sensors.

Evaporator

The evaporator was designed as a rectangular-shaped cuboid vessel, in which 4 serially-connected low-finned tubes ("MicroFin") were placed as heat exchanger (Figure 3.2). The MicroFin tubes exploit capillary action for heat transfer enhancement (Lanzerath, 2014; Lanzerath et al., 2016). The vessel was reinforced by 6 cross beams to ensure sufficient strength for vacuum conditions on the inside.

All parts were made from stainless steel except for the finned tubes which were made from copper. Further specifications on the evaporator and the finned tubes are given in Table B.2 in Appendix B. The evaporator was equipped with 2 DN40 ISO-KF flanges: one was located at the top for steam connection to the adsorber and the other on the side for a glass window to visually observe the evaporation process. Additionally, 3 DN16 ISO-KF flanges at the top and bottom of the evaporator connected the liquid reflux as well as pressure and temperature sensors. A binary filling level sensor was located between the in- and outlet (Table B.5 in Appendix B). The sensor could detect whether there was liquid or vapour phase on its tip and was used to control the evaporator filling level during operation. The sensor triggered once the evaporator was filled with approximately 300 mL of its total 700 mL volume available for the refrigerant. For precise operation of the sensor, accurate alignment of the whole evaporator in vacuum condition was crucial. The evaporator's position could be detected in two dimensions using a spirit level on 2 attached bars and was aligned by adjusting the 3 points of support.

Condenser

The condenser consisted of a cylindrical vessel, in which a plain, coiled tube was placed as heat exchanger (Figure 3.3). Steam condensed on the tube and the formed droplets freely drained along the tube towards the bottom. This way, a thin refrigerant film was avoided on the tube and the heat transfer was sufficient.

All parts except for the coiled tube were made from stainless steel. The used tubes were made from copper and were coiled up to a helix with a lathe using a special construction for tube routing. Further specifications on the condenser and the tube

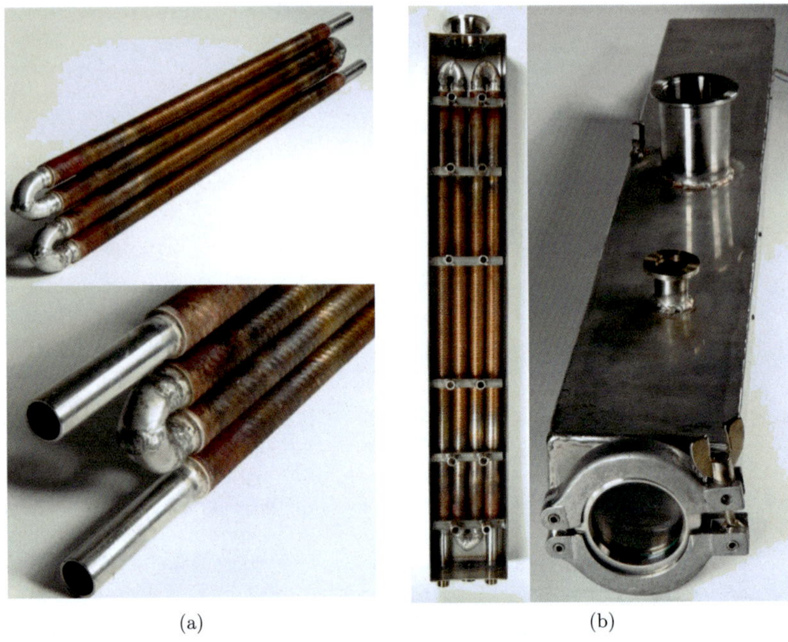

(a) (b)

Figure 3.2: Photos of the evaporator used in the adsorption chiller: heat exchanger consisting of 4 serially-connected "MicroFin" tubes in total and close-up with in- and outlet as well as "U-bend" (a) and heat exchanger in open vessel with 6 reinforcement cross beams and final evaporator with viewing window and ISO-KF flanges (b). Some ISO-KF flanges actually present on the evaporator cannot be seen on this photo.

helix are given in Table B.3 in Appendix B. The condenser was equipped with 2 DN40 ISO-KF flanges: one was located at the upper side for the steam connection to the adsorber and the other on the front at the bottom for a glass window to visually observe the condensing process and the filling level of refrigerant. Additionally, there were 2 DN16 ISO-KF flanges at the top and bottom of the evaporator to connect the liquid reflux to the glass measurement cylinder as well as pressure and temperature sensors. The bottom of the condenser was slightly tilted towards the DN16 ISO-KF flange to serve as drain. Thus, the condenser could be completely emptied, if necessary.

The glass cylinder served as refrigerant reservoir (1000 mL) and allowed for measuring the volume of the refrigerant with an uncertainty of approximately 10 mL. Hence, the volume of condensed refrigerant could be determined. Additionally, the volume of refrigerant evaporated in the evaporator could also be determined by controlling the filling level in the evaporator. Thereby, the total volume of refrigerant in the evaporator was kept constant, and the volume of refilled, liquid refrigerant from the glass cylinder to the evaporator corresponded to the volume of evaporated refrigerant.

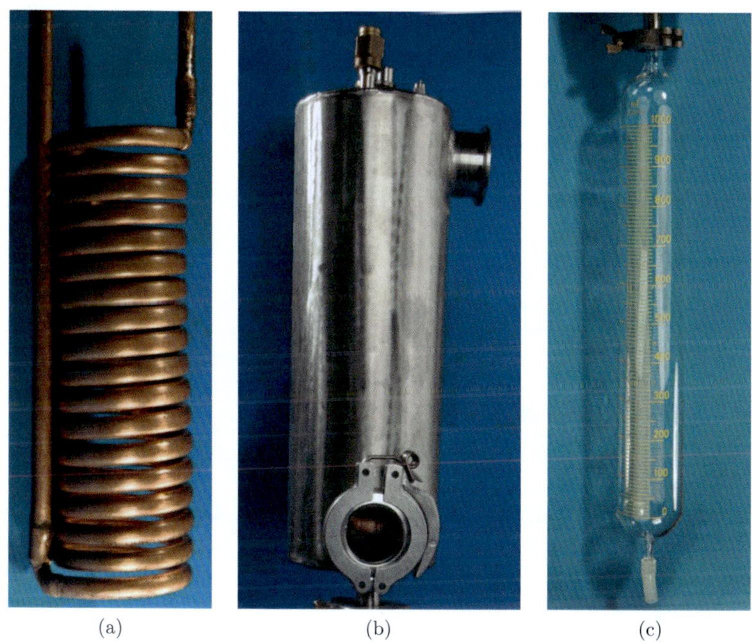

(a) (b) (c)

Figure 3.3: Photos of the condenser used in the adsorption chiller: coiled copper tube used as heat exchanger (a), condenser vessel with viewing window and ISO-KF flanges (b) and glass measurement cylinder (c).

3.1.2 Experimental setup

To operate the adsorption chiller according to the adsorption chiller cycle described in Section 2.1.1, heat flows had to be supplied to and removed from the component's heat exchangers by secondary circuits. The secondary circuit's inlet temperatures were controlled to ensure reproducible experiments. Additionally, sensors allowed to monitor the process and measure the transferred heat flows for characterisation of the adsorption chiller's performance.

Details on controlling the inlet temperature in the secondary circuits are given in the following. Specification of the experimental measurement equipment to operate the setup and determine the transferred heat flows are presented thereafter.

Inlet temperature and volume flow of secondary circuits

Thermostats controlled the inlet temperature of the heat transfer fluid for the components adsorber, evaporator and condenser. Water was used as heat transfer fluid in the secondary circuits, since it offers high specific heat capacity, good heat transfer properties and easy handling in the lab. Glykosol N (pro KÜHLSOLE GmbH, 2012) was added to the secondary circuit of the evaporator to allow for temperatures below 0 °C.

The inlet temperature of both evaporator and condenser were constant and below 100 °C. Therefore, they were directly connected to open bath thermostats which controlled the inlet temperature and volume flow (for details of the thermostats see Table 3.1).

For the operation of adsorption chillers, the adsorber inlet temperature has to be switched as described in Section 2.1.1. Additionally, the temperature level required for desorption of the adsorbent NaY is above 100 °C, thus requiring a closed, pressurised system for the heat transfer fluid water. Switching the temperature level of the secondary circuit for the adsorber was implemented with two closed thermostats and a hydraulic circuit shown in Figure 3.4 (for details of the thermostats see Table 3.1).

The used hydraulic circuit was pressurised above 10 bar with nitrogen and allowed changing the inlet temperature from 20 °C for adsorption to 160 °C for desorption within a few seconds. The volume flows for both temperature levels were realised by two pumps and could be adjusted separately by hand valves in bypasses of both pumps.

The control of the eight pneumatic inclined-seat valves (coloured in Figure 3.4) included a passive heat recovery scheme (Wang et al., 2005) by delayed switching of the outlet as shown in Table 3.2: after the valves at the inlet of the heat exchanger were switched, the hot (cold) heat transfer fluid still in the heat exchanger was continued to be pumped in the hot (cold) circuit at the outlet. The valves at the outlet of the

Table 3.1: Experimental equipment: sensors with uncertainty and thermostats with temperature stability. Locations of sensors and thermostats are given in Figures 3.4 & 3.5.

Equipment type	**manufacturer**	**model**	**uncertainty or *T*-stability**
Sensors			
Temperature	SE Sensor Electric	Pt100, class A, 4 wire connection	0.03 K
Volume flow	Huba Control	210.910441K	0.32 L min^{-1}
Mass fraction / density	Anton Paar	DMA 4500 M	
Thermostats			
Adsorber hot circuit (T_{Des})	Single	STW 200/1-12-25-H0	≤1 K
Adsorber cold circuit (T_{Ads})	Single	STW 200/1-12-36-H0.2	≤1 K
Evaporator	Haake	C41P	0.01 K
Condenser	Haake	C41P	0.01 K

heat exchangers were switched once the heat transfer fluid in the heat exchangers was completely exchanged. Thus, the phase times for the corresponding Pre-Heating and Pre-Cooling phases were chosen depending on the volume flow and volume of heat transfer fluid present in the pipes and heat exchangers between the in- and outlet valves. Successful operation of such a passive heat recovery scheme requires both identical volume and volume flow in both adsorber circuits. Nevertheless, if implemented correctly, experimental experience shows that the heat recovery scheme reduces necessary heat flows and greatly helps to ensure constant inlet temperatures for the adsorber and hence reproducible experimental conditions.

Control and experimental equipment

The experimental setup (layout depicted in Figure 3.5) allowed to operate and characterise adsorption chillers. Sensors necessary for data reduction are shown in Figure 3.5 and summarised with their standard measurement uncertainty in Table 3.1. Transferred heat flows are determined by measuring the in- and outlet temperature and

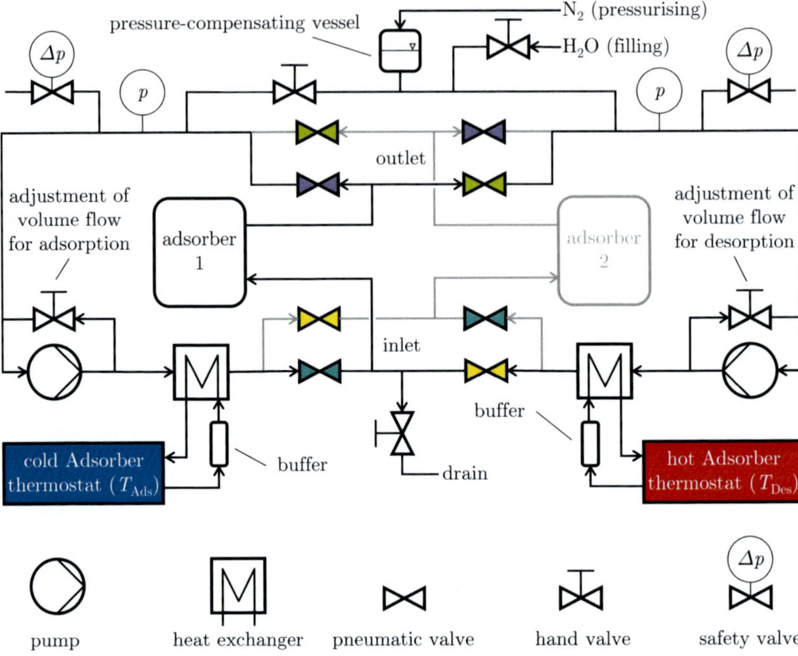

Figure 3.4: Layout of the hydraulic circuit for switching between hot and cold thermostat to control the volume flow and inlet temperature levels of two adsorbers (adsorber 2 is coloured grey, since it was not used in this work and replaced by a hose with the same volume as the heat exchanger of adsorber 1). Inlet temperatures of the adsorber secondary circuits were set by two plate heat exchangers connected to a hot resp. a cold thermostat with buffers (for details on the thermostats see Table 3.1). Switching the adsorber secondary circuits between the two heat exchangers was realised by a set of eight pneumatic inclined-seat valves. Identical coloured pneumatic valves had the same switching status, which is summarised for each phase in Table 3.2. Hand valves were used to adjust volume flow for hot and cold circuit and enable filling and draining of the circuit. The hand valve between the hot and the cold circuit was used as an expansion valve to compensate small pressure differences. The safety valves released at pressures exceeding 15 bar. Technical details of all used valves and pumps are summarised in Table B.4 in Appendix B.

Table 3.2: Switching status in the 4 phases of the adsorption chiller cycle as described in Section 2.1.1 of both the valves in hydraulic circuit for adsorber inlet temperature control shown in Figure 3.4 and butterfly / liquid valves between the adsorption chiller components evaporator (E), adsorber (A) and condenser (C). Detailed technical information about the employed valves can be found in Table B.4 in Appendix B.

Adsorption cycle phases (Figures 2.2 & 2.3)	**Inclined-seat valves in adsorber fluid circuit** (Figure 3.4) inlet		outlet		**Butterfly valves for refrigerant steam** (Figures 2.2 & 6.4)		**Liquid reflux** (Figures 2.2 & 6.4)
	⋈	⋈	⋈	⋈	E to A	A to C	C to E
Adsorption ① → ②	open	shut	open	shut	open	shut	shut
Pre-Heating ② → 2*	shut	open	open	shut	shut	shut	opened by control
Isosteric Heating 2* → ③	shut	open	shut	open	shut	shut	opened by control
Desorption ③ → ④	shut	open	shut	open	shut	open	shut
Pre-Cooling ④ → 1*	open	shut	shut	open	shut	shut	shut
Isosteric Cooling 1* → ①	open	shut	open	shut	shut	shut	opened by control

volume flow of each component. The volume flow was measured at the inlet, since the inlet temperature remains constant during the experiments. Subsequent data reduction is presented in Section 3.1.3.

The mass fraction of the additive in the refrigerant was determined at two locations: behind the condenser in the measuring cylinder and in the evaporator. Refrigerant samples were taken with an evacuated volume and two hand valves to compensate for the differences in pressure. The mass fraction of the sample was determined by density measurement (for the used density meter, see Table 3.1) and subsequent conversion to mass fraction with physical properties from Weast (1976).

The control required to operate an one-bed adsorption chiller is limited to two dimensions: inlet temperatures of the secondary circuits and length of the phase times. Inlet temperatures were chosen to be constant to allow for better comparison between

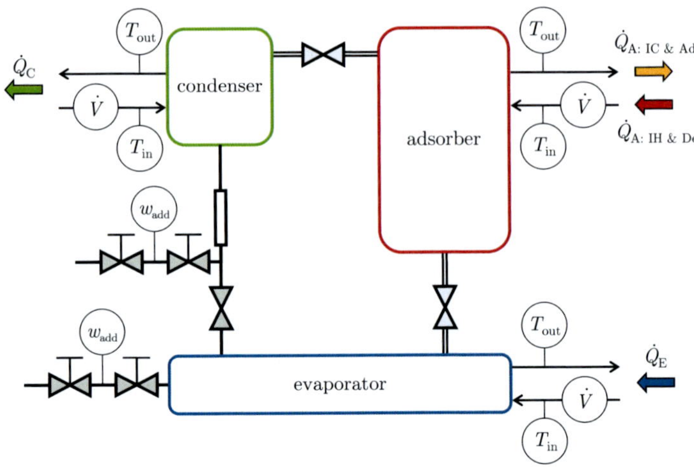

Figure 3.5: Layout of the one-bed adsorption chiller setup. Heat flows $\dot{Q}_i$ to and from all components i during the phases of the adsorption chiller cycle described in Section 2.1.1 are indicated by arrows and were measured with shown temperature T and volume flow $\dot{V}$ sensors. Additionally, sample collection for the determination of mass fraction of the additive w_{add} was possible behind the condenser and in the evaporator via vacuum locks built by two hand valves. Details on the employed sensors and their uncertainty are given in Table 3.1.

experimental studies. The phase times were imposed by changing the adsorber inlet temperature at predetermined time steps. The steam valves in the experimental setup were actively triggered, yet operated as if they were flaps reacting to the difference in pressure of the connected components. Liquid reflux from the condenser into the evaporator was controlled with a vacuum-tight direct-acting plunger valve and a peristaltic pump. Details of the used equipment are given in Table B.4 in Appendix B.

Control of the experimental setup was implemented as a timed state machine in LabVIEW. The custom-made LabVIEW program set the temperatures of the thermostats, controlled the states of the valves, read data from the sensors, applied calibrations and stored all data in a text file. The data was collected at a frequency of 1 Hz. Additionally, the front panel of the LabVIEW program displayed plots of all sensor data and valve positions necessary for monitoring the experiments. A screenshot of the front panel is shown in Figure B.5 in Appendix B.

Installed sensors included the sensors necessary for data reduction as well as additional sensors. The additional sensors included temperature sensors for component casings, liquid and steam temperatures as well as a pressure transducer in each component and supported monitoring of the process (Table B.5 in Appendix B). The pressure signals of the components were beneficial for operating the adsorption chiller. Knowledge of the pressure provided insight into the process and allowed, for example, to detect intrusion of non-condensable gases: an increasing pressure in the condenser in the course of several cycles was a very strong indication for the presence of non-condensable gases in the system as they usually accumulate in the condenser. Since the pressure in the condenser intrinsically changes due to the intermittent nature of the adsorption process, it was necessary to plot the pressure signal for the last cycles for identifying relevant increases in pressure. Non-condensable gases were removed from the adsorption chiller with a vacuum pump (Table B.4 in Appendix B).

3.1.3 Data reduction

Data collected from the in- and outlet temperature sensors ($T_{\text{in},i}$ and $T_{\text{out},i}$) and flow meter $\dot{V}_i$ were used to calculate the heat flows $\dot{Q}_i$ transferred in each component i from a steady-state energy balance on the inside of the heat exchangers:

$$\dot{Q}_i = \varrho_{\text{fluid}} c_{\text{p,fluid}} \dot{V}_i (T_{\text{in},i} - T_{\text{out},i}). \tag{3.1}$$

Here, ϱ_{fluid} was the density and $c_{\text{p,fluid}}$ the specific heat capacity of the heat transfer fluid. Water was used as the heat transfer fluid in the adsorber and condenser and its properties were calculated with TILMedia (Schulze, 2013) based on RefProp (Lemmon et al., 2013) using the equation of state "IAPWS-95" by Wagner and Pruß (2002) fitted among others to data from Cox et al. (1989); Chase (1998); Guildner et al. (1976); Kell (1975); Osborne et al. (1939) and Stimson (1969). Properties for the mixture water-Glykosol N, which was used as heat transfer fluid in the evaporator, were taken from the manufacturer's datasheet (pro KÜHLSOLE GmbH, 2012).

Once cyclic steady-state was ensured in the experiments, adsorption cycles were analysed and compared by determining the transferred heat. The transferred heats were calculated by integrating the heat flows from the start t_0 to the end t_1 of each phase of the adsorption chiller cycle (cf. Section 2.1.1):

$$Q_i = \int_{t_0}^{t_1} \dot{Q}_i \, \text{d}t. \tag{3.2}$$

3.2 Evaporation experiments

The purpose of the evaporation experiments was to determine heat transfer coefficients for sub-atmospheric, capillary-assisted evaporation on tubes designed for application in the evaporator of adsorption chillers. Specifically, the experimental setup was tailored for a study of coated tubes to investigate the impact of coating parameters on evaporation performance.

The experimental setup (Figure 3.6 (b)) consisted of two main components: the evaporator, which held the investigated tubes (Figure 3.6 (a)), and the condenser, which created the driving force for the evaporation process by lowering the pressure (Figure 3.7).

In general, the experimental setup was designed similarly to the standard experimental setup for nucleate boiling experiments suggested by Gorenflo et al. (1982), yet some differences were present: the examined heat exchangers (tubes) were not electrically heated. Instead, the required heat flow for evaporation was supplied by enthalpy flow of water on the inside of the tubes provided by thermostats (Table 3.3). Thus, effects of the inner heat transfer on the overall heat transfer coefficient of the tubes (also present in the real application) were captured in the experiment as well (cf. 3-step heat transfer in Figure 2.5). To determine the heat flow, though, the comparably simple and precise "electric" balancing of the electrical heater as proposed by Gorenflo et al. (1982) was unfortunately not possible. Instead, it was necessary to measure the difference in enthalpy flow between the in- and outlet of the evaporator.

"Determination and control of filling level" in Section 3.2.2 and Section 3.2.3 as well as Figure 3.6 and Table 3.3 of this section have been reprinted from:

J. Seiler, R. Volmer, D. Krakau, J. Pöhls, F. Ossenkopp, L. Schnabel and A. Bardow. "Capillary-assisted evaporation of water from finned tubes – Impacts of experimental setups and dynamics". *Applied Thermal Engineering* 165 (2020), 114620, with permission from Elsevier. Contributions of the author: writing the draft as one of the principal authors, adapting the design and building of the experimental setup at RWTH, planning the measurement procedure, supervising the experiments at RWTH, implementing the data reduction, and evaluating the results.

Reprinted passages use the *pluralis modestiae* "we" as an alternative to the passive voice.

Determining the difference in enthalpy flow can be accomplished by measuring the volume flow and temperature difference between the in- and outlet. Then, the heat flow for evaporation can be calculated from a steady-state energy balance on the inside of the tubes as introduced in Equation (3.1) in the data reduction for the adsorption chiller setup. However, at higher volume flows and, thus, higher heat capacity flow rates, the temperature differences can become quite small and difficult to detect when measuring small heat flows. Small heat flows can be caused, for example, by low heat transfer coefficients or a small heat transfer area, e. g. short overall tube length (cf. Equation (2.8)). Increasing the tube's length is, however, also limited: both the flow resistance and the cost for the tubes also increase with length. Additionally, the resulting temperature difference between in- and outlet should not become too large either, as the local driving force for the evaporation process becomes more inhomogeneous along the tubes. Thus, resolving physical effects caused by the local driving force becomes more difficult with longer overall tube length. Therefore, the overall length of the investigated tubes needed to be designed carefully. Here, the overall tube length was chosen to be short, yet long enough to ensure that the expected heat flows could still be measured.

To detect small heat flows from small temperature differences, accurate temperature and volume flow measurements are crucial in such a setup. The employed sensors are shown in Figure 3.6 (b) and details of the sensors are given in Table 3.3. Due to the importance of the temperature measurement, the employed temperature sensors were repeatedly calibrated and their uncertainties statistically evaluated. The process is explained in more detail in Section 3.3.1.

Accurate temperature measurement also improves the accuracy of the measured driving force for the evaporation process. The driving force was determined as the temperature difference between the temperature of the heat transfer fluid on the inside of the tube and the saturated steam temperature, which was obtained with fluid properties from the measured pressure (employed pressure transducers are also shown in Figure 3.6 (b) and details of the sensors are given in Table 3.3). Evaluating the evaporation pressure to determine the steam temperature intrinsically implies averaging of locally varying steam temperatures. Temperature sensors were also used to measure steam temperatures at several locations in the setup, yet their readings differed depending on location and time and could not be reliably reproduced. Furthermore, the presence of condensed droplets on the sensor tips resulted in unwanted artefacts and was often detected in measurement evaluation. Evaluation of the measured pressure, however, did not involve any difficulties and was, therefore, used in all experiments.

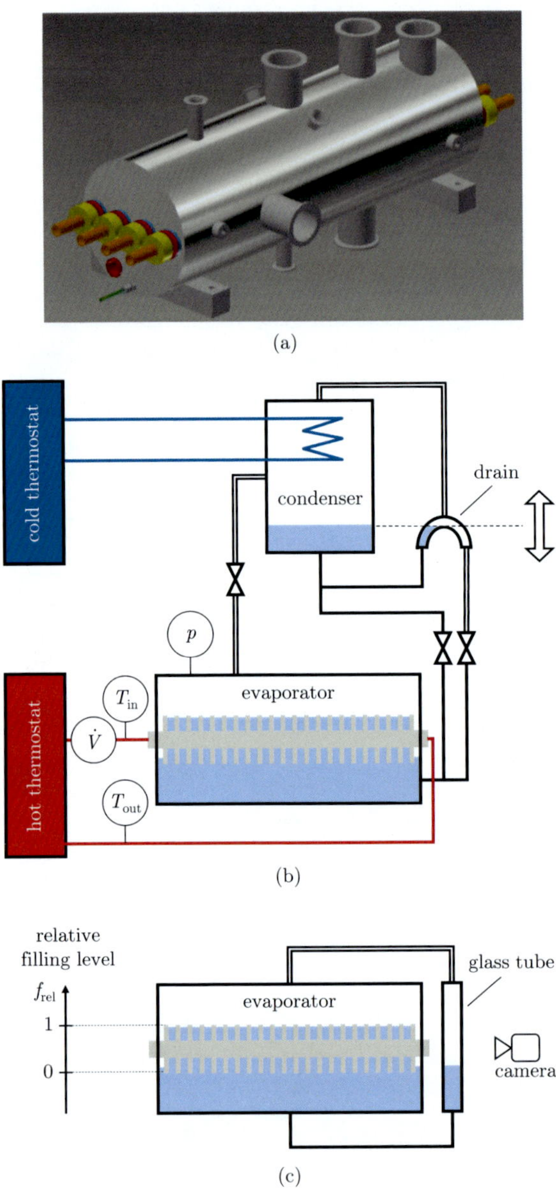
(a)
cold thermostat
condenser
drain
p
evaporator
T_{in}
$\dot{V}$
hot thermostat
T_{out}
(b)
relative
filling level
f_{rel}
1
0
evaporator
glass tube
camera
(c)

Figure 3.6: (figure on previous page) Experimental setup: (a) 3D rendered CAD drawing of evaporator chamber with 4 tubes installed. Interconnection of the 4 tubes is outside of the evaporator chamber. ISO-KF flanges are used for sensors, the vacuum pump, windows and liquid/steam connections to condenser and the communicating pipe (cf. (c)). (b) Sketch of the experimental setup consisting of finned tubes (exemplarily), the evaporator with vapour connection to the condenser and a mobile level drain for liquid reflux. The setup is shown in side view: the other three tubes lie behind the shown tube. For simplification, in- and outlet are shown on opposite sides of the first tube; in reality, they are on the same side on the first and the last of the four tubes. Capillary water between tube's fins is also shown. (c) Definition of the relative filling level f_{rel} (left) and detection of the filling level with a communicating pipe (glass tube) in the experimental setup (right).

Table 3.3: Experimental equipment: sensors with uncertainty and thermostats with temperature stability. Locations of sensors are given in Figures 3.6 (b) and 3.8.

Equipment type	**manufacturer**	**model**	**uncertainty or *T*-stability**
Sensors			
Temperature	SE Sensor Electric	Pt100, class A, 4 wire connection	absolute: 0.02 K in-out: 0.01 K
Pressure	ABB	266AST	0.04 mbar
Volume flow	Huba Control	210.910441K	$0.32\,L\,min^{-1}$
Thermostats			
Evaporator (hot)	Thermo Scientific	P1-C41P	0.01 K
Condenser (cold)	Huber	CC-510w	0.02 K

The setup (Figure 3.6) was designed as a thermosyphon: the condenser was placed above the evaporator and liquid reflux was realised by gravity (cf. Section 2.2.3). Evaporator and condenser were connected with DN40 piping for steam flow and corrugated hoses (DN15) for liquid reflux. Valves allowed to separate the components and control the liquid reflux (pneumatic DN40 steam valve and DN16 direct-acting plunger valves, see Figure 3.6 (b) for positions and Table B.9 in Appendix B for technical details). All components were thermally insulated to minimise heat flows to and from the surroundings. Figure 3.8 shows a photo of the experimental setup during assembly and provides additional information on the position of the experimental equipment.

A custom-made LabVIEW program was implemented to operate the experimental setup (Screenshots in Section B.2.3 in Appendix B). Besides acquisition, calculation, display and storing of all sensor data, the program allowed to control temperatures of the thermostats, states of the valves and auxiliary components like cameras, vacuum pump and electric heaters. Furthermore, the program was fully automated to continuously conduct experiments without staff support. Hence, time-consuming screening of experimental parameters and multiple repetitions of all experiments were possible.

For continuous, sub-atmospheric evaporation measurements, the impact of non-condensable gases inside the experimental setup is a major challenge. The presence of non-condensable gases distorts the saturation temperature measured by the pressure in the evaporator and inhibits condensation: non-condensable gases accumulate around the heat exchanger in the condenser and act as a barrier for the gaseous refrigerant. Non-condensable gases can result from many possible sources, e. g. they can be dissolved in the refrigerant or adsorbed to solid matter, form due to corrosion processes or enter the setup during the experiment due to the pressure difference to the surroundings.

Formation of non-condensable gases inside the setup by corrosion can be avoided by using corrosion-resistant materials and earthing of all components. To minimise intrusion of non-condensable gases into the setup, vacuum-tightness was a key priority in the setup's design: all components including windows and most sensors were assembled with the ISO-KF flange system. Furthermore, the setup's modular design allowed to separately test all components with a helium leak indicator to ensure vacuum-tightness step-by-step during assembly. As automation of the setup allowed to continuously operate the experimental setups for weeks, additional measures were necessary to control accumulation of non-condensable gases during operation: a small vacuum pump (Table B.11 in Appendix B) continuously purged a very small mass flow through a tiny aperture plate from the condenser. Thereby, accumulation of non-condensable gases in the setup was prevented. The effectiveness of this measure was also experimentally validated: increasing the removed mass flow did not alter the results of repeated experiments; yet when the removed mass flow was not sufficient, results were no longer reproducible.

The two main components, condenser and evaporator, are explained in detail in the following.

3.2.1 Condenser

The condenser was used to create the driving force for the evaporation process by lowering the pressure in the evaporator. To lower the pressure, refrigerant was condensed on a heat exchanger in the condenser. The heat exchanger was cooled by water with a constant inlet temperature provided by a thermostat (Table 3.3).

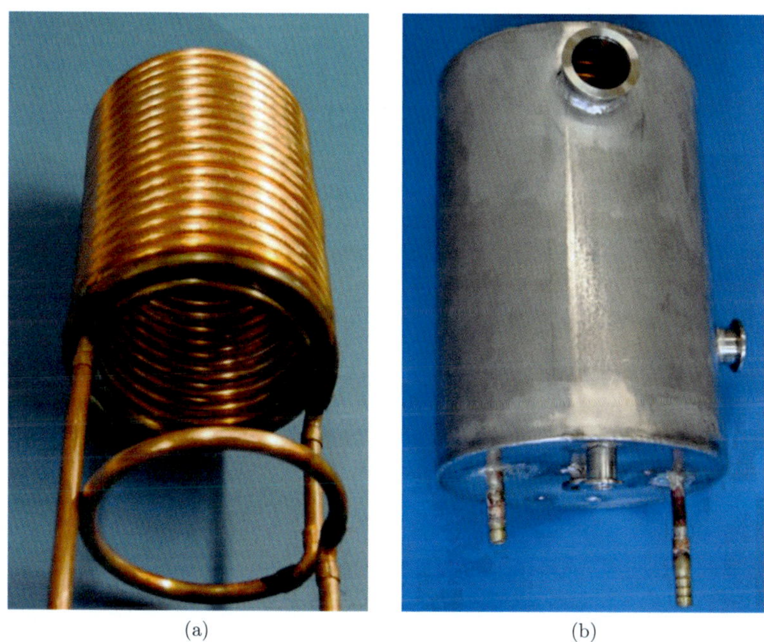

(a) (b)

Figure 3.7: Photo of the coiled-copper-tube double-helix used as heat exchanger in the condenser (a) and the condenser vessel with ISO-KF flanges and hydraulic connections for the heat exchanger (b). Geometric details of the heat exchanger, vessel and insulation are given in Table B.7 in Appendix B.

Driving force and, thus, resulting saturation pressure in the evaporator, were affected by the UA-values and inlet temperatures of the evaporator and condenser: lower inlet temperatures and higher UA-values of the condenser lead to higher driving force. However, the minimal inlet temperature of the condenser was restricted by the triple point of the refrigerant to prevent freezing in the condenser. To still allow for high driving forces, the UA-value of the heat exchanger in the condenser was chosen to be large compared to the expected UA-values of the investigated tubes in the evaporator.

The custom-made heat exchanger in the condenser (Figure 3.7 (a)) was – regarding design and function – very similar to the plain, coiled copper tube used in the adsorption chiller (cf. Section 3.1.1). However, to increase the available area for heat transfer, the coiled-copper-tube was arranged in a double-helix: an inner helix was surrounded by an outer helix separated by a small gap.

The stainless-steel vessel of the condenser (Figure 3.7 (b)) was designed as an upright standing cylinder to withstand the forces caused by the pressure difference between the ambient pressure on the outside and the low-pressure conditions present on the inside. The condenser vessel extended further than the coiled-copper-tube helixes to create a refrigerant reservoir below, which could be filled without decreasing the available UA-value for condensation. The refrigerant reservoir could store enough refrigerant to fully submerge the investigated tubes in the evaporator at least twice. The lowest turn of the copper tube in Figure 3.7 (a) was located at the bottom of the condenser to also control the temperature of the refrigerant in the reservoir.

The vessel was equipped with a DN40 ISO-KF flange on the top front for steam connection to the evaporator and 4 additional DN16 ISO-KF flanges: one flange was located at the side, covered with a glass window to provide a view of the inside, one was located at the top to attach the vacuum pump and an additional pressure transducer and 2 flanges were located at the bottom. The 2 DN16 flanges at the bottom were used for liquid reflux: one was located at the lowest point of the bevelled bottom, allowing to fully drain the condenser and was usually sealed with a blind flange. The second flange extended a little further into the inside of the condenser to prevent complete draining of the refrigerant reservoir by leaving a liquid pool in the condenser at the bottom. Since the temperature of the remaining liquid pool was controlled by the extra turn of the heat exchanger, the pressure in the condenser was also continuously controlled by the temperature of the heat exchanger. Without the remaining pool, the pressure in the condenser was not constant, resulting in non-uniform driving forces.

Geometric details of the heat exchanger, vessel and insulation are given in Table B.7 in Appendix B.

3.2.2 Evaporator

The evaporator vessel was designed to determine the heat transfer coefficients of tubes, which exploit capillary action for efficient sub-atmospheric evaporation. These tubes resembled a heat pipe, albeit turned inside out: heat pipes are usually sealed pipes which feature capillary action on their inside surface to transport liquid refrigerant from the condenser to the evaporator section (cf. Section 2.2.3). The investigated tubes provide capillary action on their outer surface to transport liquid refrigerant, creating a thin film of refrigerant for efficient evaporation.

The enthalpy of vaporisation was supplied by flow of water on the inside of the tubes. The tubes were surrounded by the evaporator vessel, a vacuum-tight, horizontal cylinder made from stainless steel (CAD drawing in Figure 3.6 and picture in Figure 3.8). The geometric details of the evaporator vessel, insulation and suitable tube dimensions are given in Table B.6 in Appendix B.

The evaporator vessel was filled with liquid refrigerant. The filling level of the evaporator determined how far the tubes were submerged in the refrigerant. Since the filling level is an important characteristic of capillary-assisted thin-film evaporation (cf. Section 2.2.5), its determination and control is explained in more detail in Section 3.2.2.

Alignment of the four 4 tubes among each other and alignment compared to the refrigerant's surface (spirit level) are crucial to uniformly submerge the tubes. To achieve the best possible alignment, manufacturing of the evaporator vessel was sophisticated and included milling of the mounting for the vacuum fittings on both sides of the welded vessel with a special machine in one setting. Final alignment of the tubes with a spirit level was realised by adjusting the evaporator's position with fine-threaded screws in the ground plate. Since the vessel was designed as a cylinder, even evacuating the vessel did not alter the alignment – a problem that was present in former design iterations of the experimental setup.

The evaporator always contained 4 identical tubes. The tubes were serially connected on the outside of the evaporator vessel. Thus, only the parts of the tubes providing capillary action were located inside the vessel and contributed to evaporation. The connections of the tubes on the outside did not participate in heat transfer and, additionally, were not required to be vacuum-tight. Usually, vacuum-tightness needs to be ensured by soldering the tubes; here outside of the vacuum vessel, simple connections with hose clamps were sufficient and, additionally, allowed for quick exchange of the tubes (cf. Figure 3.9).

Vacuum-tight sealing between the tubes and the evaporator vessel was achieved by specialised vacuum fittings, in which O-rings were sealed by knurled nuts – allowing for quick exchange of the tubes by manual assembly (Figure 3.9). The corresponding sealing surface on the tubes was faced on a lathe to enable proper tightness.

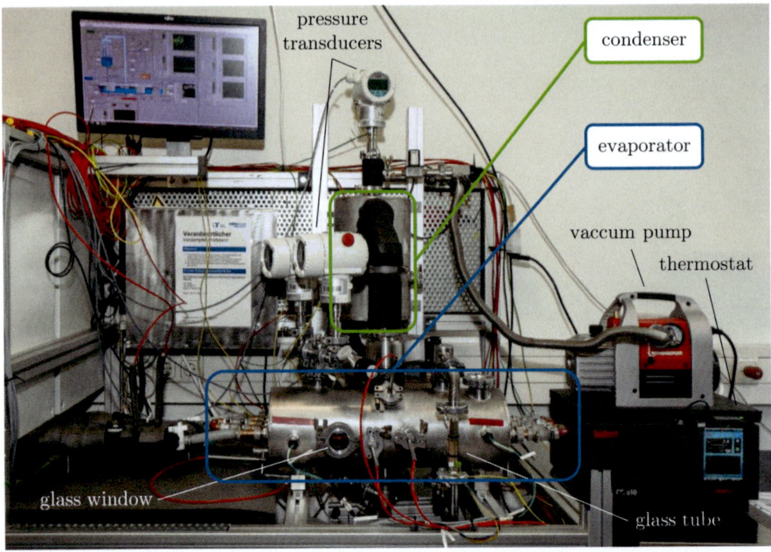

Figure 3.8: Photo of the setup for the evaporation experiments during assembly. The evaporator and condenser as well as the external parts of the tubes (extending on both sides of the evaporator) are not yet insulated. In- and outlet of the evaporator are on the left-hand side: the inlet is connected to the tube in the front. Thus, the glass window to the right of the inlet allows viewing the presumably hottest part of all tubes. Another glass window is located on the top on the right-hand side of the evaporator and offers a view on the top of the investigated tubes. The 3 sensors with digital displays and white casings are pressure transducers. All red cables lead to Pt100 temperature sensors, which are installed through insulated metal canning tubes serving as antikink devices and insulation. A glass tube to show the filling level in the evaporator is installed up front with a flexible, corrugated hose to allow for thermal expansion without breakage. The thermostat setting the condenser's inlet temperature and the vacuum pump are located on the right-hand side. The hot thermostat, which is connected to the tubes, cannot be seen on the picture, since it is placed below the shown setup.

An additional advantage of such a vacuum-tight sealing is the fact that the tubes are also better thermally insulated from the vessel: the only possibility of heat transfer is through the refrigerant or the O-ring and not from the copper tube directly into the stainless-steel vessel.

Exchange of the tubes required venting of the evaporator vessel and, consequently,

Figure 3.9: Photo of the vacuum-tight sealing between the tubes and the evaporator vessel. Quick exchange of the tubes was possible by manually operated fittings, which sealed O-rings by knurled nuts, as well as connection of the tubes with hose clamps.

thorough subsequent evacuation to minimise the effect of non-condensable gases on the experiments (cf. above). To reduce the chance of non-condensable gases dissolving in the refrigerant, the refrigerant was usually stored in the condenser without contact to air during the exchange of the tubes. Nevertheless, degassing the experimental setup was necessary each time, and experience showed that evaporation of the refrigerant was a fast and effective way of degassing. To enable fast evaporation of all liquid refrigerant in the evaporator, 2 additional electrical cartridge heaters were installed in a stainless-steel tube at the bottom of the evaporator vessel. An automated degassing program guaranteed quick degassing of the experimental setup.

Determination and control of filling level

As the condenser was located above the evaporator, gravity was used for the refill process of liquid refrigerant into the evaporator (Figure 3.6 (b)). The evaporator was directly flooded with the liquid refrigerant up to the desired filling level. The filling level could either decrease during the evaporation experiments by omitting any refilling and keeping the liquid refrigerant in the condenser or the filling level could be kept constant by continuous reflux of the liquid refrigerant from the condenser.

To allow for measurements at constant filling level, the experimental setup was equipped with a drain (Figure 3.6 (b), not shown in Figure 3.8). Once the refrigerant in the condenser reached the level of the drain, any additional condensed refrigerant flowed directly back into the evaporator and the filling level in the evaporator was kept constant. The filling level in the evaporator could be decreased – without venting the setup (!) – by moving the level drain upwards such that more refrigerant was stored in the condenser before the refrigerant reached the level drain and flowed into the evaporator. Correspondingly, by moving the level drain downwards, the filling level in the evaporator was increased.

In this thesis, the filling level of the evaporator is given as the relative filling level f_{rel} which equals 1 if the filling level is at the top of the tube and 0 if it is at the bottom of the tube as shown in Figure 3.6 (c, on the left) and in Figure 2.13 in a sectional view. Negative filling levels are possible when the filling level is below the tube but surface tension still maintains a connection between the tube and the water surface by a meniscus.

Detection of the relative filling level f_{rel} in the evaporator proved difficult with standard measuring techniques based on detection of density, capacitance, speed of sound, etc., since these properties strongly depend on temperature for the considered conditions close to the triple point of water and due to low reliability of available sensors (e. g. effect of water droplets). Instead, we developed an optical system to determine the relative filling level f_{rel} of the refrigerant (Figure 3.6 (c), right).

The optical system consisted of a communicating pipe to show the filling level in a glass tube outside of the evaporator chamber and a digital camera (cf. Table B.8 in Appendix B) to capture images of the visible phase interface. The location of the phase interface was determined by a self-programmed edge-detection algorithm (cf. Figure 3.10). First, the binarisation step of the algorithm used separate thresholds for each rgb plane of the picture to obtain a black and white picture of the meniscus. Since suitable threshold values highly depended on the lighting conditions, white LEDs were used to provide uniform lighting. Next, the algorithm detected the position of the lower edge of the refrigerant meniscus on the glass tube. The final inversion step of the algorithm is not necessary for the edge-detection, but simplifies visual identification of the detected meniscus on a computer screen to check the results. The position of the lower edge was evaluated relative to the millimetre scale of the attached graph

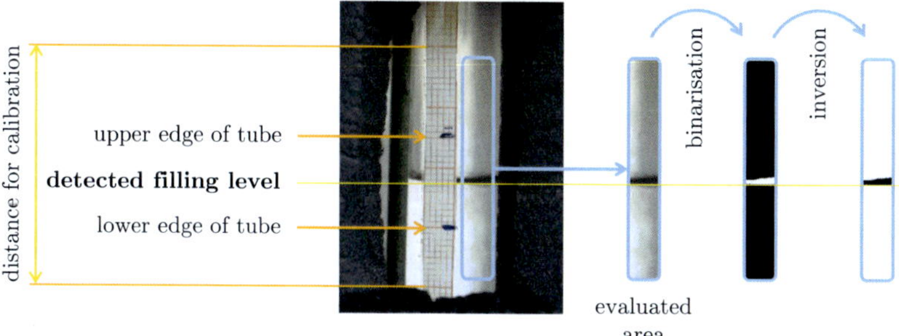

Figure 3.10: Optical detection of the evaporator filling level from pictures of the glass tube acting as a communicating pipe (cf. Figure 3.8 and Figure 3.6 (b)). Steps of the self-programmed algorithm to evaluate the digital pictures and automatically detect the **filling level** are also shown. For reference, the filling levels corresponding to the upper and lower edge of the coated tubes (cf. Chapter 5) are also indicated.

paper to determine the relative filling level. Automatic evaluation of 300 pictures per experiment resulted in reproducible measurement of the relative filling level f_{rel} (cf. Figures D.7 and D.8 in Appendix D), which was verified by additional visual evaluation of pictures from each experiment.

3.2.3 Data reduction

To characterise the heat transfer from the heat transfer fluid through the tubes to the refrigerant vapour phase, we used the overall heat transfer coefficient U

$$U = \frac{\dot{Q}}{A \Delta T_{\mathrm{ln}}}, \tag{3.3}$$

where $\dot{Q}$ is the heat flow from the heat transfer fluid through the reference surface area A of the tubes into the refrigerant on the outside of the tubes. We chose the logarithmic mean temperature difference ΔT_{ln} to describe the driving force for this heat flow,

To determine the heat flow $\dot{Q}$, we applied a steady-state energy balance around the tubes

$$\dot{Q} = \dot{V}\varrho(h_{\text{in}}(T_{\text{in}}) - h_{\text{out}}(T_{\text{out}})), \tag{3.4}$$

where $\dot{V}$ is the volume flow of the heat transfer fluid inside the tubes, ϱ the density, $h_{\text{in}}(T_{\text{in}})$ and $h_{\text{out}}(T_{\text{out}})$ the specific enthalpies of the heat transfer fluid at the in- and outlet of the evaporator tubes. Heat flows from the surroundings into the evaporator were compensated for by calibrating the outlet temperature: the required offset was obtained by measuring the difference between in- and outlet temperature in steady-state conditions for the closed evaporator chamber, i.e., every time before the evaporation process started. The determined offsets were of the same order of magnitude as the uncertainty of measurement.

The reference surface area A was chosen as the cylindrical surface enclosing the fins to account for the necessary construction volume of the finned tubes. Thus, the diameter d_{o} (from fin tip to fin tip) and the length L of the investigated n tubes are included in the calculation of A:

$$A = n\pi d_{\text{o}} L. \tag{3.5}$$

The logarithmic mean temperature difference ΔT_{ln} is the driving force and was determined by

$$\Delta T_{\text{ln}} = \frac{T_{\text{in}} - T_{\text{out}}}{\ln\left(\frac{T_{\text{in}} - T_{\text{v}}}{T_{\text{out}} - T_{\text{v}}}\right)}. \tag{3.6}$$

Here, T_{in}, T_{out} and T_{v} are the fluid temperatures at the evaporator in-/outlet and the vapour temperature inside the evaporator chamber, respectively. Since direct measurements of the vapour temperature are difficult, we used the saturation temperature of water corresponding to the measured pressure p in the evaporator chamber (assuming absence of non-condensable gases inside the chamber and thermodynamic equilibrium in the vapour phase):

$$T_{\text{v}} = T_{\text{sat}}(p). \tag{3.7}$$

As for the adsorption experiments (Section 3.1), all needed fluid properties were calculated with TILMedia (Schulze, 2013) based on RefProp (Lemmon et al., 2013) using the equation of state "IAPWS-95" by Wagner and Pruß (2002) fitted among others to data from Cox et al. (1989); Chase (1998); Guildner et al. (1976); Kell (1975); Osborne et al. (1939) and Stimson (1969).

3.3 Analysis of measurement uncertainty

In general, all measurements are subject to errors, i. e. a difference remains between the measured value and the unknown, true value. Unfortunately, the true value can usually not be determined. Instead the **uncertainty** of a measurement is given as an interval around the measured value together with the underlying probability that the true value lies within that interval. Hence, when comparing two different measurements, not only the measured value itself, but also the interval describing the uncertainty of the measurement should be evaluated. Therefore, knowledge of the uncertainty of a measurement is crucial to assess the significance of the measurement result itself and of the derived conclusions.

The commonly accepted framework to assess measurement uncertainty in standard technical measurement tasks is called "Guide to the expression of uncertainty in measurement" (GUM) (Joint Committee for Guides in Metrology, 2008, 2009). The GUM describes how to (I) estimate standard uncertainties for measurements, (II) combine the standard uncertainties of different measurements and (III) convert combined uncertainties with a coverage factor to adjust the probability that the true value lies within the reported interval. In this thesis, all uncertainties are given with a coverage factor of $k = 1$, leading to reported intervals having a level of confidence of approximately 68 %. More details on the application of the GUM to assess the measurement uncertainty in this thesis can be found in Appendix C.

To assess the uncertainty of the whole measurement chain, sensors can and should be calibrated while already being connected to the data acquisition of the experimental setup whenever possible. The calibration of sensors used in this thesis and subsequent assessment of measurement uncertainty according to GUM is discussed in the next section.

3.3.1 Sensor calibration: correction of measured values and quantification of uncertainty

Calibration is defined as the comparison of a measured value with the measured value of a calibrated measurement device (i. e. its uncertainty is known) for the same measurement task. A comparison of the two values obtained allows to assess the difference and, if repeated, conclusions about the uncertainty of the determined difference.

The experimental setups in this thesis use 3 types of sensors to measure relevant values used in data reduction: pressure transducers, volume flow meters and temperature sensors.

The calibration of the pressure transducers as well as the volume flow meters were conducted by the manufacturer and the sensor's uncertainties are stated in the respective calibration certificates. To check the consistency of the stated uncertainties, measured values of the pressure transducers and volume flow meters were compared for the same measurement task: in pretests, the volume flow meters showed the same readings and in case of the pressure transducers, two identical sensors were simultaneously operated in the experimental setup during all measurements. Both pressure transducers continued to show identical readings within the measurement uncertainty during all conducted experiments. Thus, the consistency checks of the uncertainty stated by the manufactures were successful.

Temperature sensors are usually sold without calibration certificates as the uncertainty of measurement strongly depends on the data acquisition hardware. Instead, temperature sensors belong to a certain "precision grade" according to a standard set by Deutsches Institut für Normung e.V. (2009). The temperature sensors used for this thesis are Pt100, stainless steel, 1.5 mm in diameter, class A, with 4 wire connection by the manufacturer SE Sensor Electric and their maximum deviation at temperature T in °C is according to class A

$$\pm\left(0.15+0.002\,|T|\right)\ \mathrm{K}. \tag{3.8}$$

The readings of the used Pt100 temperature sensors were compared to 2 calibrated temperature sensors (Ahlborn, Pt100 type FPA923L0250, connected to Ahlborn Almemo 1030-2, standard uncertainty for temperatures from 0 to 100 °C according to DAkkS-certificate: 0.0075 K) in a calibration experiment: all temperature sensors were connected to the data acquisition of the experimental setup and inserted in matching holes of a copper block placed in a thermal bath (Figure 3.11). The evaluation started once all sensor readings were settled, i. e. the overall difference of each sensor's readings during the last 100 s was below 0.01 K. Since both calibration sensors were placed in holes located at different diameters of the cylindrical copper-block, it can be assumed that the temperature of the copper block was homogenous (within the uncertainty of the calibration sensors) and all temperature sensors were exposed to the same temperature.

The deviations of each sensor to the mean temperature T_{AB} of the calibration sensors A & B were then averaged for a time span of 100 s. Afterwards, the procedure was

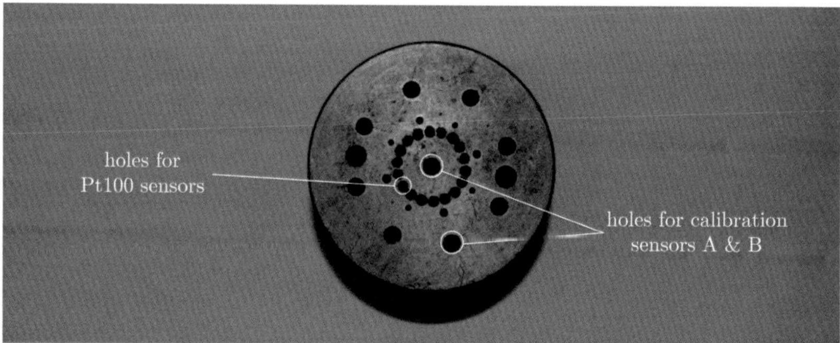

Figure 3.11: Custom-made copper block with holes of different diameters used for calibration of temperature sensors with different thicknesses. The calibration sensors A & B were inserted in the holes in the middle and at the farthest diameter of the cylinder to be able to assess the temperature distribution of the symmetrical block. Pt100 sensors were inserted into holes in between. For calibration, the whole block, including the sensors, was submerged in thermal oil in a bath of a thermostat.

repeated with the next temperature step. The temperature steps were chosen according to the required temperature range of the temperature sensors. Exemplarily, determined deviations for 10 temperature sensors from −5 °C to 80 °C in 5 K increments are given in Figure 3.12.

Since the whole experiment was automated, repetitions of the experiment for statistical analysis could be conducted without additional manpower. In Figure 3.12, the average deviations including the experimental standard deviations are given for 6 repetitions.

Additionally, the maximum allowed deviation according to the "precision grade" class A (Deutsches Institut für Normung e.V., 2009) of the employed sensors is plotted in Figure 3.12. As can be seen, most sensor's deviations lie within the specified deviation and all deviations follow a parabolic curve, which is typical for deviations of Pt100 sensors. However, the determined experimental standard deviations of the repeated experiments indicate that the uncertainty of the examined temperature sensors can be drastically reduced if the determined deviations are corrected.

Such a correction was implemented using the experimentally determined deviations and linear interpolation between the measured points. The evaluated sensors differ both regarding their deviation from the calibration sensors A & B and regarding their

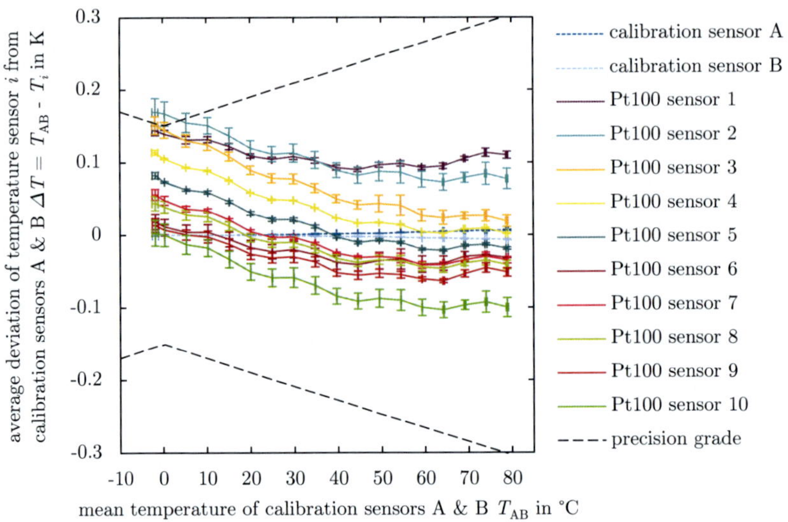

Figure 3.12: Average deviations during calibration of temperature sensors Pt100 including standard experimental deviation for 6 repetitions of the experiment. Also shown are the deviations of the 2 calibrated temperature sensors A & B as well as the maximum allowed deviation of the Pt100 sensors according to the "precision grade" class A (Deutsches Institut für Normung e.V., 2009).

determined experimental standard deviation. Sensors with smaller standard deviations can be regarded as more suitable for higher reproducibility. Thus, sensors with small experimental standard deviations were chosen to be employed to measure temperatures crucial for data reduction. Furthermore, temperature sensors, which mainly measure temperature differences (e. g. in- and outlet), were chosen with a preferably similar course of the determined deviations.

According to GUM, statistical analysis of repeated experiments allows estimating the standard measurement uncertainty. Following Equations C.4 and C.2 in Appendix C, the standard measurement uncertainty of the temperature sensors was determined from the experimental standard deviation from Figure 3.12. The resulting standard measurement uncertainties are shown in Figure 3.13. All Pt100 sensors show a standard measurement uncertainty below 0.02 K for all evaluated temperatures (except sensors 2 and 10 at temperatures below 0 °C). Some sensors (1, 4, 5 and 7) actually

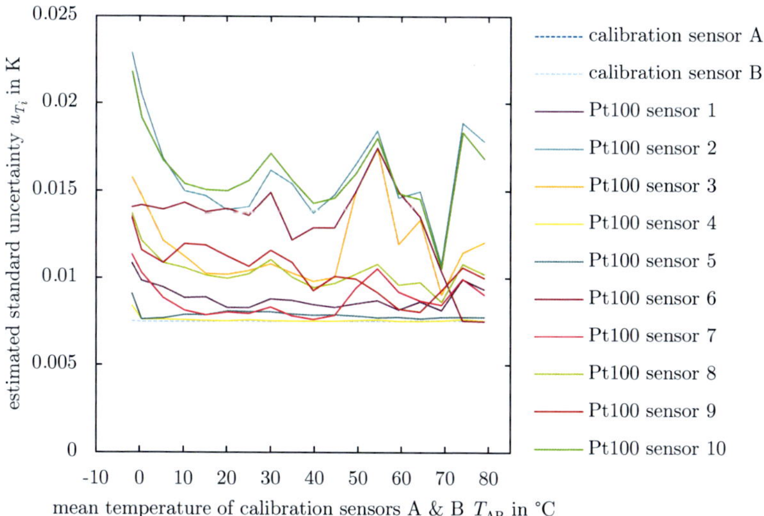

Figure 3.13: Standard measurement uncertainty of temperature sensors Pt100 determined from standard experimental deviation in Figure 3.12.

mostly stay below 0.01 K, which demonstrates the quality of the employed temperature sensors and calibration procedure. Nevertheless, the uncertainty for further analysis of measurement uncertainty in data reduction was chosen conservatively to be 0.02 K (cf. Table 3.3) to account for systematic contributions to uncertainty that cannot be accounted for in the described procedure.

Experimental experience shows that measuring temperatures with a measurement uncertainty of 10 mK is challenging and many factors need to be taken into consideration to accomplish the presented results (e. g. electromagnetic fields from other devices, fluctuations or oscillations in the power grid, earthing of all electrical conducting parts of the experimental setup, etc.). Furthermore, characteristics of the sensor's installation and the liquid's flow condition may also affect the temperature measurement, e. g. even in turbulent flows, the temperature distribution within the liquid flow might not necessarily be homogenous within 10 mK.

Since direct calibration of the temperature sensors installed in the experimental setup is impossible, the calibration of the installed temperature sensors was checked by an additional hydraulic construction. Temperatures measured at the in- and outlet of the heat exchangers can be compared, when the heat exchanger is bypassed and the temperature of the heat exchanger fluid is evaluated by the sensors in very close distance. If the distance is close enough and insulated, temperature changes of the heat exchanger fluid due to exchange of heat with the surroundings are very small and can be neglected. Such a situation can be ensured by installing 2 manual 3/2 directional control valves (Table B.9 in Appendix B) between the in- and outlet of the component close to the temperature sensors (Figure 3.14). The valves have a low thermal mass (e. g. for the low-temperature circuits the material was polyvinylidene fluoride) and allow to quickly bypass the heat exchanger to compare the readings of the in- and outlet temperature sensors.

The readings of the in- and outlet temperature sensors of all components were repeatedly compared in an automated experiment, which was very similar to the calibration experiment for the absolute temperature described above: the experiment was conducted at equidistant temperature steps to cover the relevant temperature range. The experimental procedure was identical, although the temperature readings (including the corrections from the absolute calibration experiment) were not compared to the calibration sensors. Instead, the differences of the in- and outlet temperature sensors were evaluated.

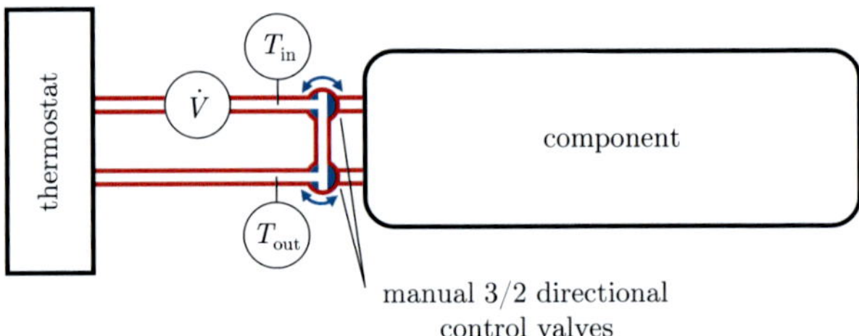

Figure 3.14: Layout for relative calibration employed in all components. Manual 3/2 directional control valves located close to the temperature sensors allow to bypass the component for calibration of the temperature sensors relative to each other. The bypass is insulated and – if not in use – emptied to minimise unwanted heat flows during both calibration and regular operation.

The determined mean differences from all 5 repetitions of this "relative calibration" always stayed below the determined standard measurement uncertainty of the temperature sensors (Figure 3.13). Furthermore, even the maximum measured differences in each measurement step of all repetitions stayed below the determined uncertainty, highlighting the reproducibility of the temperature measurement.

Since the determined differences were reproducible with an experimental standard deviation of below 1/10 of the determined standard measurement uncertainty of the employed temperature sensors (Figure 3.13), the differences were also corrected and the measurement uncertainty for the temperature difference $T_{in} - T_{out}$ for further analysis of measurement uncertainty in data reduction was conservatively estimated to be 0.01 K (cf. Table 3.3).

Such a low measurement uncertainty of the measured temperature differences seems reasonable, since, for relevant sensors, the determined standard measurement uncertainty (Figure 3.13) was already below 0.01 K which was additionally confirmed by the relative calibration.

In summary, the described hydraulic design offers the advantage to verify the relative calibration of the in- and outlet temperature sensors while being installed in the experimental setup. Additionally, the verification can be conducted regularly, since the experimental effort is limited to operating the manual valves and starting an automated sequence.

Chapter 4

Validating setup and procedure for the evaporation experiments

Reproducibility of experiments is crucial for reliable results. Results need to be reproducible in repetitions of the same experiment in the same setup, yet their reliability is even further increased if they are also reproducible in another setup.

Contents of this chapter (Section 4.1 to Section 4.4) have in slightly amended form been reprinted from:

J. Seiler[a], R. Volmer[b], D. Krakau[a], J. Pöhls[b], F. Ossenkopp[b], L. Schnabel[b] and A. Bardow[a,c]. "Capillary-assisted evaporation of water from finned tubes – Impacts of experimental setups and dynamics". *Applied Thermal Engineering* 165 (2020), 114620, with permission from Elsevier. Contributions of the author: writing the draft as one of the principal authors, adapting the design and building of the experimental setup at RWTH, planning the measurement procedure, supervising the experiments at RWTH, implementing the data reduction, and evaluating the results.

[a] RWTH Aachen University, Institute of Technical Thermodynamics, Germany
[b] Fraunhofer Institute for Solar Energy Systems ISE, Department Heating and Cooling Technologies, Germany
[c] Forschungszentrum Jülich, Institute of Energy and Climate Research, Energy Systems Engineering (IEK-10), Germany

Reprinted passages use the *pluralis modestiae* "we" as an alternative to the passive voice.

In this chapter, the experimental setup and procedure used for the evaporation experiments in this thesis (cf. Chapter 5) are validated by comparing the results for evaporation on identical tubes in this setup to another experimental setup. The two employed experimental setups were located at RWTH Aachen University[a] (this thesis) and Fraunhofer ISE[b] and the conducted experiments were part of a collaboration between research groups at both institutions.

Both research groups employed their experimental setup for experiments on sub-atmospheric, thin-film evaporation of water before and published results on this topic prior to the collaboration, e. g. Lanzerath et al. (2016), Seiler et al. (2019) and Volmer et al. (2017).

4.1 Introduction

Unfortunately, many results on sub-atmospheric, thin-film evaporation of water reported in literature cannot be directly compared due to differences in heat exchanger type and experimental parameters, such as (I) evaporation temperature, (II) driving force and (III) filling level. Furthermore, the impact of the employed experimental setups and procedures has – to the best of our knowledge – not been analysed for sub-atmospheric thin-film evaporation of water. Our measurement experience suggests that also the "hidden" factors linked to experimental setup and procedure can potentially have a large impact on the experimental results – but can easily be missed in measurement evaluation: e.g., non-condensable gases intruding the vacuum system can influence condensation (Wang and Tu, 1988) and evaporation (Crößmann, 2016) and distort vapour pressure measurements. Consequently, the validity, repeatability and comparability of published measurement results can often not be ensured today.

Therefore, the first objective of this chapter is to fill this gap by identifying and discussing potential impacts due to setups and procedures on thin-film evaporation measurements. On that basis, we want to disclose which steps are necessary to realise comparable and reproducible evaporator characterisation. For this purpose, we investigate a set of identical tubes at the same experimental parameters in two different experimental setups. As the main evaluation quantity, we use the overall heat transfer coefficient U.

In both setups, measurements are conducted with decreasing filling level since this type of measurement offers the advantage of including all possible filling levels in one single experiment. However, these dynamic experiments create only quasi-steady-state conditions whose impact on the evaporation process has not been determined yet.

Therefore, as a second objective, we assess whether experiments with decreasing filling levels can be compared to experiments with constant filling levels.

We used two different experimental setups located at two different research institutions: one experimental setup was located at the Institute of Technical Thermodynamics at RWTH Aachen University, Germany (from now on referred to as RWTH) and the other experimental setup was located at the Fraunhofer Institute for Solar Energy Systems ISE in Freiburg, Germany (from now on referred to as ISE). The experimental setups were operated by the authors according to their affiliation stated above.

From our experimental findings, we deduce methodological recommendations to contribute to good scientific practice for improving reliability and comparability of measurement results for sub-atmospheric thin-film evaporation of water.

4.2 Materials and methods

4.2.1 Investigated finned copper tubes

All evaporation measurements were carried out with the natural refrigerant water (R-718) on the same type of finned copper tubes, provided by Wieland-Werke AG, Germany. Geometric characteristics of the tubes were chosen in a way that a strong capillary effect and, hence, high evaporation performance can be expected at the given conditions. An internal structure ensured a high heat transfer on the fluid side. Thus, changes in the overall heat transfer coefficient U can be attributed to thin-film evaporation on the refrigerant side. Geometric specifications of the finned tube are listed in Table 4.1 and a photo of a finned tube installed in the experimental setup at RWTH is given in Figure 4.1.

The heat exchangers consisted of four identical tubes, which were serially connected (cf. Figures 3.6 and 4.2). Each tube section had a finned length of 0.5 m which yielded a total length of 2 m. The distance between the tube axes was 39 mm. Since measurements were conducted in two different experimental setups, the tube heat exchangers were configured in a slightly different way: for the RWTH setup, the tubes were led through the evaporator chamber walls and interconnected outside the evaporator chamber (Figure 3.6 (a) and (b)). For the ISE setup, "U" bends were soldered to the tube ends and the whole heat exchanger was installed and connected to the fluid circuit inside the evaporator chamber (Figure 4.2).

Table 4.1: Geometric specifications of the finned tube.

Material	Tube inner diameter	Tube outer diameter (fin tips)	Fin spacing	Fin width	Fin height
CU (C12200)	9.6 mm	13.2 mm	56 fpi	0.15 mm	0.8 mm

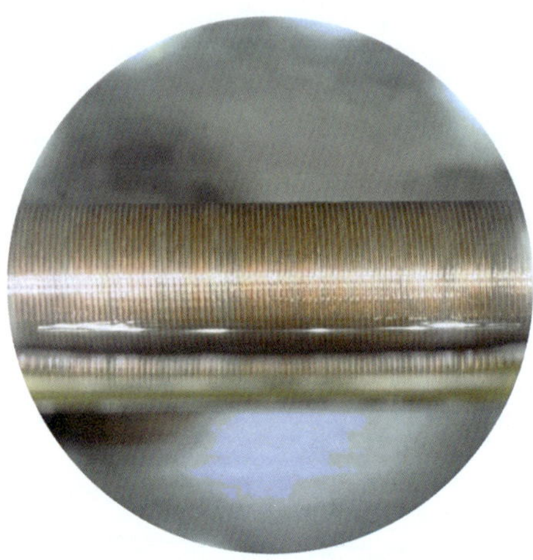

Figure 4.1: Finned tube installed in the experimental setup at RWTH partly submerged in the refrigerant water during an experiment. The picture is taken through a round window on the side of the experimental setup. The other three tubes lie behind the shown tube. Capillary water between the fins of the tube as well as the menisci of water at the tube and at the window can be seen.

4.2.2 Experimental setups

The two experimental setups at RWTH and at ISE had a similar design concept but differed in some aspects: Both setups consisted of a vacuum-tight system to allow for sub-atmospheric pressure evaporation of the refrigerant water. All materials were chosen such that corrosion-induced formation of non-condensable gases inside the system was avoided. Both setups featured a vacuum chamber, in which evaporation on the finned tubes took place, and a condenser which created the driving force for the evaporation process and which represented a refrigerant reservoir. In the fluid circuits for the evaporator and condenser, two thermostats were used to set the volume flow and inlet temperature of the heat transfer fluid water.

The specifics of the RWTH setup are described in Section 3.2 while the ISE setup is presented in the following.

Experimental setup ISE

In the experimental setup at ISE (Figure 4.2), the evaporator chamber integrated hydraulic lead-throughs which allowed installation of various kinds of heat exchangers via vacuum flange connections inside the chamber. The condenser consisted of an over-sized flat tube-fin heat exchanger which was permanently welded into a rectangular vacuum chamber. The condenser heat exchanger was always partially flooded with refrigerant and, thus, served as a refrigerant reservoir while always maintaining sufficient free surface area for condensation. The chambers for evaporation and condensation were connected by two vapour lines which could be opened or closed by automated valves.

Since the condenser is placed below the evaporator, refrigerant cannot flow from the reservoir in the condenser to the evaporator chamber by gravity. Instead, refrigerant is supplied to the evaporator chamber by a refrigerant pump (not shown in Figure 4.2 (b)) or by condensation on the investigated evaporator. This configuration allows for operational modes in which excess refrigerant in the evaporator is drained back to the condenser by gravity. Evaporation can be measured both in steady-state and transiently. The measurements described in this work were carried out in transient mode and refrigerant was supplied to the test heat exchanger by condensation.

To obtain comparable time series of measured quantities, it was necessary to ensure that evaporation of a certain amount of refrigerant lead to the same change in filling level in both experimental setups: different filling levels would alter the heat transfer rate and, thus, the time series of all measured quantities. This requirement of equal filling levels can be realised by creating the same base area in both refrigerant pools. For this purpose, at ISE, an open transparent acrylic glass (PMMA) trough with roughly the same base area as the refrigerant pool area at RWTH ($0.0812\,\mathrm{m}^2$) was placed inside the evaporation chamber (Figure 4.2 (a)). Thus, not the whole evaporator chamber was flooded with refrigerant but only the trough. The finned-tube heat exchanger was placed inside the PMMA trough propped on two support bars of 10 mm height. Only the "U"-bends were in contact with the bars so that the finned tube sections were not blocked. Thus, the refrigerant filling level could drop below the bottom edge of the fins (Figure 4.2 (b)), allowing for relative filling levels below 0.

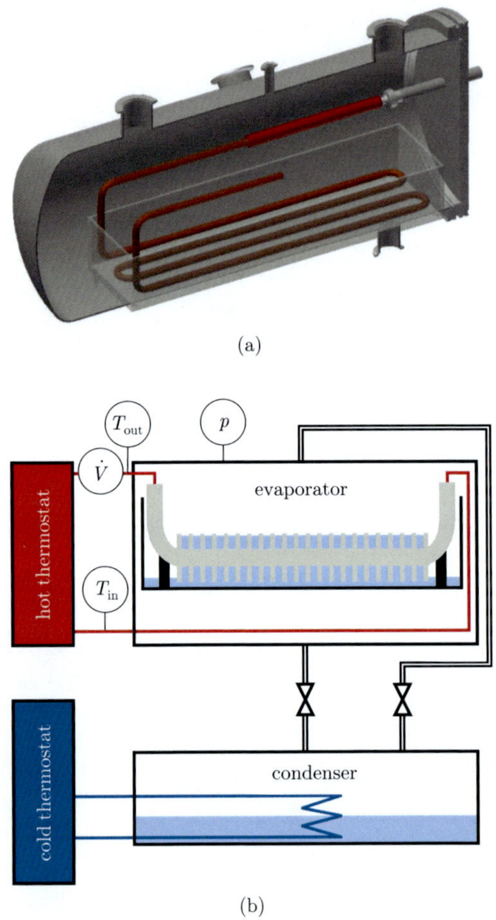
(a)
T_{out}
p
$\dot{V}$
evaporator
hot thermostat
T_{in}
condenser
cold thermostat
(b)

Figure 4.2: (figure on previous page) Experimental setup at ISE: (a) CAD drawing of the heat exchanger and trough inside half-section of evaporator chamber. Interconnection of the 4 tubes is inside the evaporator chamber. ISO-KF flanges are used for sensors, the vacuum pump, windows and vapour connections to the condenser. (b) Sketch of experimental setup consisting of investigated finned-tube evaporator on two support bars (coloured black) in acrylic glass (PMMA) trough placed in an evaporator chamber with vapour connection to the condenser below. The setup is shown in side view: the other three tubes lie behind the shown tube. For simplification, in- and outlet are shown on opposite sides of the first tube, In reality, they are on the same side on the first and the last of the four tubes. Capillary water between fins is also shown.

4.2.3 Data reduction

Both experimental setups measured the same quantities: volume flow $\dot{V}$, in- and outlet temperature T_{in} and T_{out} of the heat transfer fluid inside the evaporator tubes and the pressure p in the evaporator chamber (cf. Figures 3.6 (b) and 4.2 (b)). Thus, we could employ the same data reduction in both cases (Section 3.2.3). Details on the used sensors and thermostats are given in Tables 3.3 and 4.2. For the flowmeters and the pressure transducers, we used the calibration provided by the manufacturer. The temperature sensors were calibrated while being connected to the data acquisition of the experimental setups.

Table 4.2: Experimental equipment in setup at ISE: sensors with uncertainty and thermostats with temperature stability. Locations of sensors are given in Figure 4.2.

Equipment type	manu-facturer	model	uncertainty or T-stability
Sensors			
Temperature	tmg GmbH	Pt100, class 1/3B, 4 wire connection	absolute and in-out: 0.011 K
Pressure	MKS	Baratron 627B	0.043 mbar
Volume flow	Krohne	Optiflux 1010C/D+IFC 010 D	0.033 L min^{-1}
Thermostats			
Evaporator (hot)	Huber	CC-515w	0.02 K
Condenser (cold)	Huber	CC-515w	0.02 K

4.2.4 Uncertainty in measurement

We calculated measurement uncertainty according to the "Guide to the expression of uncertainty in measurement" (GUM) using a coverage factor of $k = 1$ (Joint Committee for Guides in Metrology, 2008) as described in Section 3.3.

For our assessment of uncertainty, we accounted for the whole measurement chain including uncertainty of the sensors and data acquisition hardware. Uncertainty in the maximum measured U-value is below 13 % for all experiments. Uncertainties for all calculated quantities are given in Figures D.3 to D.6 in Appendix D. The uncertainty of the U-value is quite high for low heat flows and low driving forces, particularly due to the uncertainty of the temperature and pressure measurement. Therefore, it is crucial to use temperature and pressure transducers with low measurement uncertainty for this type of experiment.

To assess the consistency of the results from both experimental setups (RWTH and ISE), we use the coefficient of variation CV as metric

$$\mathrm{CV} = \frac{\sqrt{\frac{1}{t_1 - t_0} \int_{t_0}^{t_1} \left(y_{\mathrm{RWTH}} - y_{\mathrm{ISE}}\right)^2 \, \mathrm{d}t}}{\frac{1}{2}\left(\frac{1}{t_1 - t_0} \sum_{i=t_0}^{t_1} y_{\mathrm{RWTH}}(i) + \frac{1}{t_1 - t_0} \sum_{i=t_0}^{t_1} y_{\mathrm{ISE}}(i)\right)}. \tag{4.1}$$

The coefficient of variation CV relates the root mean square deviation to the mean of the compared quantity from both setups y_{RWTH} and y_{ISE} from timestep t_0 to timestep t_1. A CV close to zero shows that the results of both experiments are in good agreement.

4.2.5 Experimental procedure

Prior to starting the measurement series, the experimental setups were brought to a well-defined state to ensure comparable measurement conditions at both labs (RWTH and ISE). The heat exchangers were cleaned with isopropanol to remove grease from the production process and installed inside the evaporator chamber. The evacuated refrigerant reservoirs were sufficiently filled with degassed, deionized/bi-distilled water. After filling, remaining non-condensable gases were thoroughly removed from all system components with a vacuum pump (cf. Section D.3 in Appendix D). Additionally, for each evaporation measurement, non-condensable gases were again removed: in the ISE setup, the condenser and evaporator were evacuated for 4 minutes each, prior to the evaporation. In contrast, in the RWTH setup, the condenser was evacuated through a small aperture plate during the whole experiment to allow for continuous operation of the setup.

Before each evaporation measurement, the heat exchanger tubes were completely flooded with refrigerant so that all surfaces were wetted. Afterwards, all valves connecting the evaporator chamber and the condenser were closed. Subsequently, the thermostats were set to the target temperatures of the evaporator and condenser fluid inlets. When all components had reached steady-state conditions, the valves (one valve at RWTH, two valves at ISE) between evaporator and condenser chamber were opened to start the evaporation process. In the course of the measurement, the filling level in the evaporator decreased due to evaporation. When the filling level reached the upper edge of the finned tubes, the partially flooded phase started where thin refrigerant films form on the tubes' surfaces due to capillary action in the fin interstices.

For measurements with decreasing filling level, the filling level continuously decreased until the measurements ended when the bottom edges of the tubes lost contact to the refrigerant surface.

For steady-state measurements with constant filling level – which were only conducted at RWTH – the level drain (see Figure 3.6 (b)) was used to keep the filling level constant at the desired levels. After all measured quantities had reached a steady state (after approximately 30 minutes), the evaporation process was continued for 2 minutes and the quantities measured during these 2 minutes were averaged. Afterwards, the level drain was moved upwards to decrease the filling level to the next constant filling level.

4.2.6 Input parameters

To compare measurements at RWTH and ISE, input conditions of the experiments need to be identical. The input conditions are: evaporator inlet temperature, volume flow and pressure. Thermostats were used to ensure constant and identical volume flow as well as inlet temperature to the evaporator (Table 4.3). The temperature set both the driving force of the evaporation process and the evaporation temperature (which is associated with the evaporator pressure).

Ideally, the evaporation pressure should also be constant. However, keeping the evaporation pressure constant would require a rather complex control strategy of the experimental setup since the fluid inlet temperature of the condenser would have to be controlled dynamically by the thermostat. Such a control strategy was not implemented at either of the two locations. Therefore, the condenser inlet conditions were set to obtain comparable evaporation pressures in both experimental setups. As the evaporation pressure was not constant during the experiments, the minimum and maximum values are given in Table 4.3.

Table 4.3: Input parameters for the comparison of experimental setups at RWTH and ISE.

Parameter set	**low Driving Force: lowDF**		**high Driving Force: highDF**	
Experimental setup	RWTH	ISE	RWTH	ISE
Evaporator inlet temperature	15 °C	15 °C	15 °C	15 °C
Evaporator volume flow	7.9 L min^{-1}	7.9 L min^{-1}	7.9 L min^{-1}	7.9 L min^{-1}
Evaporator inlet *Re*	1.53×10^4	1.53×10^4	1.53×10^4	1.53×10^4
Min. evaporator pressure	15.7 mbar	15.7 mbar	13.3 mbar	13.1 mbar
Max. evaporator pressure	16.4 mbar	16.4 mbar	15.1 mbar	15.1 mbar
Filling level	continuously decreasing	continuously decreasing	continuously decreasing	continuously decreasing

To investigate the impact of the driving force, measurements were performed for two pressure ranges in the evaporator with the same fluid inlet temperature leading to an experiment with low driving force (lowDF) and an experiment with high driving force (highDF). The corresponding settings are given in Table 4.3.

Each experiment was repeated at least once in each setup (RWTH and ISE) to guarantee reproducibility of the results.

The steady-state measurements were carried out at 6 constant filling levels f_{rel} which covered a range from approximately 0.75 to −0.3. For comparison, measurements were also conducted with decreasing filling level. Further settings for these measurements are given in Table 4.4. Each measurement was performed at least three times to guarantee reproducibility.

Table 4.4: Input parameters for comparison of continuously decreasing filling level with constant filling levels.

Parameter set	Decreasing filling level	Constant filling level
Experimental setup	RWTH	RWTH
Evaporator inlet temperature	15 °C	15 °C
Evaporator volume flow	$11.0\,\mathrm{L\,min^{-1}}$	$11.0\,\mathrm{L\,min^{-1}}$
Evaporator inlet *Re*	2.13×10^4	2.13×10^4
Min. evaporator pressure	13.1 mbar	14.1 mbar
Max. evaporator pressure	15.1 mbar	15.1 mbar
Filling level	continuously decreasing	constant at 6 different levels

4.3 Results and discussion

4.3.1 Impact of experimental setups

The discussion is divided in three parts: comparison of the input quantities, description of the dynamics of the output quantities and comparison of the output quantities.

Comparison of input quantities

Measured input conditions of experiments are shown in Appendix D.1 (Figures D.1 and D.2) for both measurement parameter sets and experimental setups. During thin-film evaporation until the contact between the tubes and the water is lost, the input conditions of both setups show good agreement within uncertainty of measurement. Evaporator volume flow and inlet temperature were constant and almost identical. The evaporator pressure was not constant during the experiments since no control strategy for keeping the pressure constant was implemented. Instead, the condenser input conditions were kept constant. Thus, the evaporator pressure dynamically adapted to the changing U-values of the evaporator and condenser. Still, the resulting shapes and absolute values of the evaporator pressure curve coincide within the uncertainty of measurement for both setups. Thus, the prerequisite for comparing the measurements is fulfilled.

Dynamics of output quantities

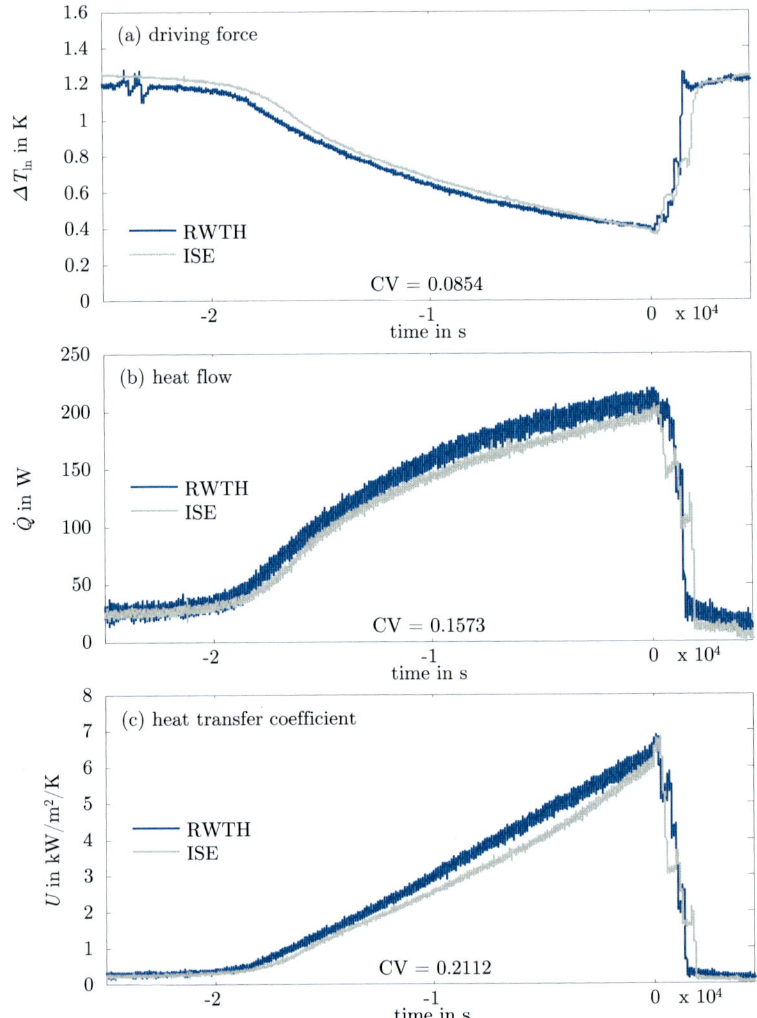

Figure 4.3: (figure on previous page) Driving force ΔT_{ln} (a), heat flow $\dot{Q}$ (b) and overall heat transfer coefficient U (c) plotted over time for experiments with parameter set lowDF (Table 4.3). Both experiments (RWTH and ISE) are synchronized at the end of the experiment (time=0) when the first tube loses contact to the refrigerant pool. The coefficient of variation CV (Equation (4.1)) is given in each plot to quantify the deviation between the experiments at RWTH and ISE.

Resulting values for driving force ΔT_{ln}, heat flow $\dot{Q}$ and overall heat transfer coefficient U are nearly constant at the beginning of the experiment (flooded phase) (Figures 4.3 and 4.4). When thin-film evaporation starts (partially flooded phase), the driving force ΔT_{ln} at first decreases strongly (heat flow $\dot{Q}$ increases strongly); with decreasing filling level, the curves become less steep. The reason for these dynamics is as follows: in the flooded phase, the thermal resistance and available surface for evaporation are nearly constant; during thin-film evaporation, the thermal resistance becomes smaller and the area covered with thin-film gradually increases which increases the evaporation heat flow $\dot{Q}$. Due to the circular shape of the tube's cross section, the gain in thin-film surface increases fast in the first phase of thin-film evaporation while the gain becomes smaller when the filling level approaches the half height of the tube. As evaporation power increases, the system pressure p and, thus, the saturation temperature increase while the evaporator outlet temperature decreases which results in a decrease of the driving force ΔT_{ln}. Decreasing slope of evaporation heat flow $\dot{Q}$ and increasing slope of driving force ΔT_{ln} mostly compensate each other in the calculation of the U-value which results in a virtually linear increase of the U-value over time. The linear increase in U-value is mainly due to an increase in available surface area for thin-film evaporation, as the reference surface area A of the tubes used to calculate the U-value (cf. Equation (3.3)) is constant and does not change according to the actually available thin-film surface area for the heat transfer (Lanzerath et al., 2016).

Thus, the maximal U-value occurs at the end of the experiment, shortly before the contact between the tubes and the refrigerant pool is lost. The detachment of the refrigerant pool from the tube happens at relative filling levels f_{rel} well below 0 since a meniscus continues to connect the tubes to the refrigerant pool even if the pool level is below the tube. The detachment occurs separately for all four tubes which can be seen in four distinctive drops of the U-value at the end of the experiment. Shortly before each drop in U-value, the U-value strongly increases. This increase in U-value is caused by an increase of available surface area: the area which was previously occupied by the meniscus connecting the refrigerant pool to the tube is now made available for thin-film evaporation by the detachment.

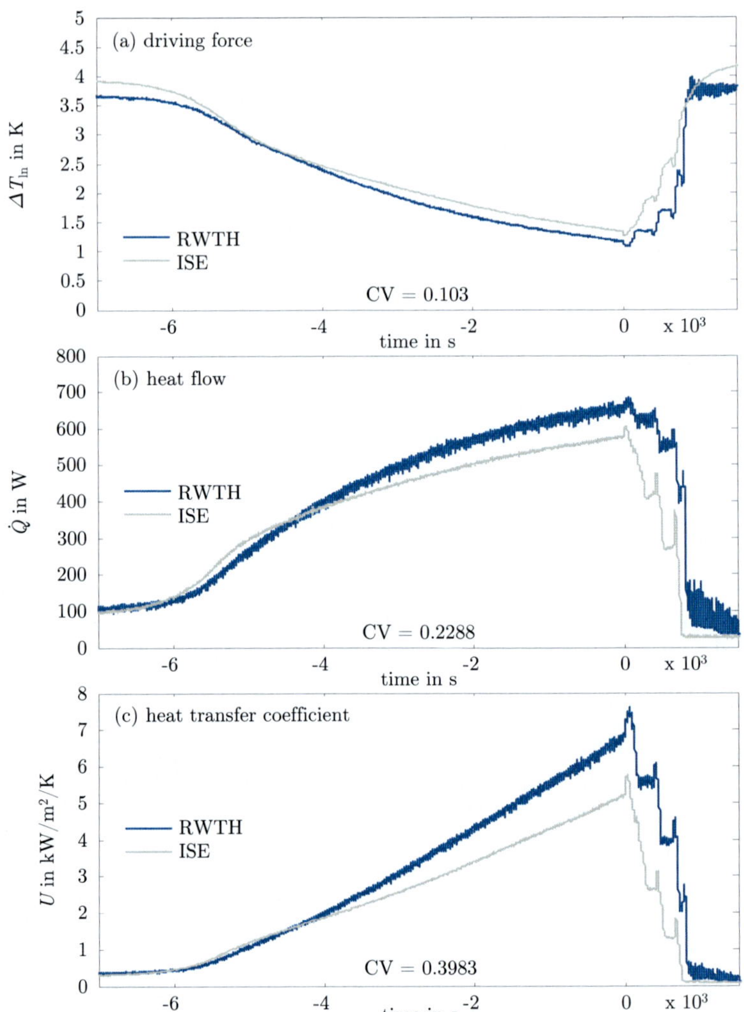
(a) driving force
$\Delta T_{\ln}$ in K
RWTH
ISE
CV = 0.103
time in s
x 10³
(b) heat flow
$\dot{Q}$ in W
RWTH
ISE
CV = 0.2288
time in s
x 10³
(c) heat transfer coefficient
U in kW/m²/K
RWTH
ISE
CV = 0.3983
time in s
x 10³

Figure 4.4: (figure on previous page) Driving force ΔT_{ln} (a), heat flow $\dot{Q}$ (b) and overall heat transfer coefficient U (c) plotted over time for experiments with parameter set highDF (Table 4.3). Both experiments (RWTH and ISE) are synchronized at the end of the experiment (time=0) when the first tube loses contact to the refrigerant pool. The coefficient of variation CV (Equation (4.1)) is given in each plot to quantify the deviation between the experiments at RWTH and ISE.

The comparison of different experiments should, however, be based on the maximum U-values measured before the contact to the refrigerant pool is lost, since this operating point can reliably be detected. In practice, relative filling levels f_{rel} above 0 will most likely be chosen for continuous operation to ensure a reliable connection between refrigerant and the tube.

Measured maximum U-values for parameter sets lowDF and highDF are quite high in both experimental setups compared to values reported in literature: maximum measured U-values range from approximately $5.2\,\text{kW}\,\text{m}^{-2}\,\text{K}^{-1}$ to $6.8\,\text{kW}\,\text{m}^{-2}\,\text{K}^{-1}$ for driving forces ΔT_{ln} from 0.4 K to 1.3 K (Figures 4.3 and 4.4).

The maximum overall heat transfer coefficients reported in literature for comparable thin-film evaporation heat exchangers at similar operating conditions are slightly lower: Seiler et al. (2019) (i. e. Chapter 5) and Lanzerath et al. (2016) report values of approx. $5\,\text{kW}\,\text{m}^{-2}\,\text{K}^{-1}$. Most likely, the higher U-values in the current work are due to enhanced inner heat transfer of the tubes. Xia et al. (2008) investigated tubes with a larger diameter, though the dimensions of the outer structures are similar to the ones examined here. At 15 °C evaporator inlet temperature, approx. 2 K driving force and a relative filling level of 0.5, their reported heat transfer coefficients on the refrigerant side (4 to $5.2\,\text{kW}\,\text{m}^{-2}\,\text{K}^{-1}$) are of the same order of magnitude as the overall heat transfer coefficients (U-values) reported here (3 to $5\,\text{kW}\,\text{m}^{-2}\,\text{K}^{-1}$). Since internal heat transfer is not limiting in our case, heat transfer coefficients on the refrigerant side are probably only slightly higher than the overall heat transfer coefficients which allows a rough comparison. Thimmaiah et al. (2016) also investigated thin-film evaporation on similar tubes, yet they reported overall heat transfer coefficients (U-values) below $1\,\text{kW}\,\text{m}^{-2}\,\text{K}^{-1}$ at evaporator inlet temperature of 15 °C. In our view, reasons for the lower values are most likely a limitation of the heat transfer on the inner side of the tube and differences in the experimental determination of the driving force ΔT.

Comparison of output quantities

For parameter set lowDF (Table 4.3), measurements at RWTH and ISE show overall good agreement (Figure 4.3): the curve shapes of the driving force ΔT_{ln}, heat flow

$\dot{Q}$ and U-value for RWTH and ISE are very similar and the values coincide within uncertainty of measurement (Figures D.3 and D.4 in Appendix D).

The driving force ΔT_{ln} (Figure 4.3 (a)) determined at ISE is most of the time slightly higher than at RWTH. Consequently, the calculated CV is larger than the maximum CV calculated for the inlet conditions (< 0.005, Figure D.1 in Appendix D); nevertheless, the correspondence is excellent.

The measured heat flows $\dot{Q}$ (Figure 4.3 (b)) are almost identical during natural convection of the submerged tubes at the beginning of the experiment. However, starting with the onset of thin-film evaporation, the heat flow $\dot{Q}$ measured at RWTH is slightly higher than at ISE. Furthermore, the random noise is more pronounced in the experimental setup at RWTH (Figures 4.3 and 4.4) which is due to stronger fluctuations in the evaporator inlet conditions (Figures D.1 and D.2 in Appendix D) and higher uncertainty of the flowmeter at RWTH. The deviations between RWTH and ISE are reflected in a higher CV of the heat flows $\dot{Q}$ compared to the CV of the driving force ΔT_{ln}.

The U-values of thin-film evaporation determined at RWTH in Figure 4.3 (c) are slightly higher throughout the experiment. However, the U-values coincide within uncertainty of measurement with the U-values obtained in the setup at ISE (Figure D.4 in Appendix D). Maximum and minimum U-values are almost identical (deviation $< 4\,\%$). The CV increases compared to the CV of the heat flows and logarithmic temperature difference, but is still comparably low.

The experiments with parameter set lowDF show that comparable results at different experimental setups can be obtained for thin-film evaporation experiments. For parameter set lowDF, the influence of the experimental setup seems to be small, since the determined values coincide well within uncertainty of measurement.

For parameter set highDF (Table 4.3), measurements are again similar but show overall less agreement than for parameter set lowDF (Figure 4.4): the driving force ΔT_{ln} and heat flow $\dot{Q}$ deviate between RWTH and ISE, though the U-value still appears linear for both setups.

The driving forces ΔT_{ln} at detachment resulting from parameter set highDF are considerably larger than for parameter set lowDF: 1.25 K vs. 0.4 K. The minimum and maximum values of the driving force ΔT_{ln} are in good agreement (Figure 4.4 (a)) although they do not coincide within the uncertainty of measurement (Figures D.5 and D.6 in Appendix D). Thus, the CV is larger than the maximum CV of the input conditions (< 0.02, Figure D.2 in Appendix D) and larger than the respective CV for parameter set lowDF.

The heat flows $\dot{Q}$ increase compared to parameter set lowDF (maximum values in the range of 600 W vs. 200 W) due to higher driving forces ΔT_{ln}. As for parameter set lowDF, the minimum values of the heat flow $\dot{Q}$ during natural convection of submerged tubes are in good agreement between both setups. However, the heat flow $\dot{Q}$ increases less at ISE compared to RWTH during thin-film evaporation. Thus, the maximum heat flows $\dot{Q}$ differ: 650 W (RWTH) vs. 580 W (ISE). The deviation of about 11 % is not within the uncertainty of measurement and increases the CV compared to the CV of the driving force ΔT_{ln} and results for parameter set lowDF.

The U-values for natural convection of the submerged tubes are also in good agreement. However, the U-value increases more strongly over time at RWTH than at ISE. As a consequence, the maximum U-values differ: 6.8 kW m^{-2} K^{-1} (RWTH) vs. 5.2 kW m^{-2} K^{-1} (ISE). The deviation of 24 % is not within uncertainty of measurement and is reflected in a strongly increased CV compared to the CV of the driving force ΔT_{ln} and heat flow $\dot{Q}$ as well as the respective results for parameter set lowDF.

Comparing the results of the measured U-values at RWTH and ISE for the parameter sets lowDF and highDF, it becomes apparent that at low driving force (parameter set lowDF) the results match very well while at higher driving force (parameter set highDF) the results deviate to a greater extent. Furthermore, the measured U-values at RWTH do not show a significant dependence on driving force (maximum U-values 6.5 and 6.8 kW m^{-2} K^{-1} for parameter sets lowDF and highDF) whereas ISE results do suggest such a dependence (maximum U-values 6.1 and 5.2 kW m^{-2} K^{-1} for parameter sets lowDF and highDF). This outcome indicates that the presence of a driving force impact might depend on the experimental conditions.

In the following paragraphs, the three most plausible explanations for the observed deviations between RWTH and ISE results are discussed. Further potential influencing factors – which are considered as generally important but not for our case – are pointed out in detail in Appendix D.3.

(I) **Surface properties.** Literature has shown that thin-film evaporation strongly depends on the surface properties of the heat exchanger: surface morphology (e. g. roughness), chemistry (e. g. oxidation state), physics (e. g. polarisation) and contamination influence wettability (de Gennes et al., 2009; Yekta-fard and Ponter, 1985; Shirazy et al., 2012) which is crucial for evaporation (Lee et al., 2014; Yuki et al., 2015; Kim and Kang, 2003). Prior to our measurements, we tried to provide identical surface conditions by removing contaminants with a cleaning process using isopropanol. However, soldering of the "U" bond connectors to the finned tubes might have involuntarily modified the surface of the heat exchanger for the ISE setup, e. g. by oxidation due to exposure to high temperatures or contamination with solder or flux agent residuals. Also, exposure to different surrounding media during storage and installation (water vapour, air, gaseous contaminants) might have changed surface conditions. Since thin-film evaporation can be very sensitive to wetting behaviour, we consider differences in surface conditions as a very likely reason for the differences in the measured heat transfer coefficients: possibly, wettability of the ISE heat exchanger was deteriorated by the soldering process to such an extent that capillary forces did no longer suffice to completely fill the space between the fins at lower filling levels and a partial dry-out occurred. This hypothesis would match the observation that U-values from RWTH and ISE deviate only at higher driving forces where partial dry-out of the tube surface becomes more likely due to higher evaporation mass flow.

(II) **Configuration of refrigerant pool.** At RWTH, the refrigerant pool has a large volume and it is in direct contact with the walls of the evaporator chamber, while at ISE, the volume is smaller and it is located inside the acrylic glass trough inside the chamber. Thus, at RWTH, the pool has a higher heat capacity and could gain more heat from the surroundings. These heat gains do not affect the energy balance since a correction for heat gains was conducted as described in Section 3.2.3. However, due to lower heat gains and heat capacity at ISE, evaporation from the pool surface supposedly leads to lower temperatures of the refrigerant pool at ISE compared to RWTH. Thus, the physical properties of the refrigerant are different which could decrease the mass flow through capillary action and therefore explain the lower U-values.

(III) **Measurement uncertainties.** For analysing measurement uncertainties, we used accuracy specifications of the manufacturers' datasheets for the pressure transducers and flowmeters and calibrated the temperature sensors. We also included uncertainties of the data-acquisition systems (see Tables 3.3 and 4.2 as well as Section 4.2.4). However, there might be unforeseen contributions to uncertainty or systematic effects (such as drift effects) which are not included in the uncertainty analysis. Consequently, the discrepancies between the measurement results from the two different experimental setups might to some extent be due to unknown uncertainties and/or distorting effects.

4.3.2 Impact of dynamic effects: constant vs. continuously decreasing filling level

In the experiments for the comparison of the experimental setups (Figures 4.3 and 4.4), the filling level continuously decreased since the refrigerant evaporated but no liquid flowed back from the condenser to the evaporator (Figures D.7 and D.8 in Appendix D). Thus, each experiment allowed to evaluate all filling levels ($1.2 > f_{\text{rel}} > -0.25$). However, the advantage of evaluating all filling levels within one experiment comes at the cost of not having truly steady-state conditions during the experiment. To assess the difference of these "quasi-steady-state conditions" (Lanzerath et al., 2016) compared to true steady-state conditions, we conducted additional experiments using the mobile level drain for liquid reflux (Figure 3.6 (b)) to create constant filling levels in the experimental setup at RWTH.

We conducted experiments with evaporator inlet conditions according to Table 4.4 and a condenser inlet temperature of 10 °C with both continuously decreasing filling level and 6 constant filling levels. Both types of experiments were repeated at least 3 times and the measured U-values are given in Figure 4.5.

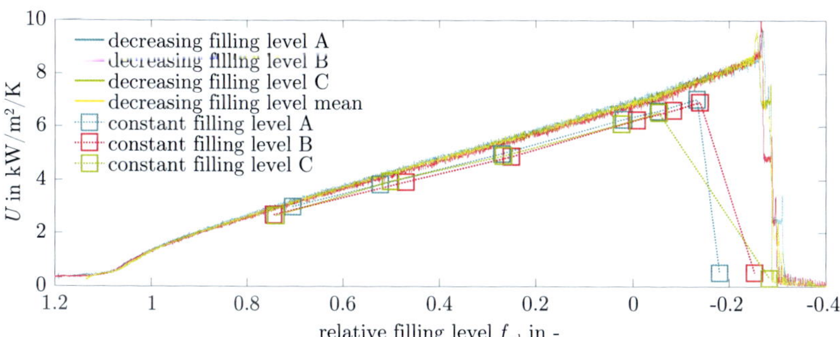

Figure 4.5: U-values for experiments with continuously decreasing filling level (for each repetition A-C and arithmetic mean of A-C) and constant filling level (for each repetition A-C) plotted over relative filling level f_{rel}. Operating conditions of evaporator were according to Table 4.4. Condenser temperature was set to 10 °C. Data points of experiments with constant filling levels are connected by a dashed line to mark subsequently recorded data points.

In Figure 4.5, the repetitions of both types of experiments are in very good agreement. Furthermore, both types of experiments show the same trend. Therefore, experiments with continuously decreasing filling levels are well-suited for a quick assessment of the U-value at all filling levels.

However, all experiments with continuously decreasing filling levels lose contact to the refrigerant at lower filling levels and show generally 5-10 % higher U-values than the experiments with constant filling levels. Although the differences are close to or within the uncertainty of measurement, the good reproducibility of the experiments indicates that there are underlying systematic reasons for the decrease in U-value. Possible reasons can be:

(I) **Differences in measurement conditions due to measurement uncertainty.** Since the observed differences for experiments with continuously decreasing and constant filling levels are close to or within the uncertainty of measurement, ensuring identical inlet conditions is crucial. For the shown experiments, the inlet conditions are identical within the uncertainty of measurement, yet still not exactly the same. However, the remaining differences cannot be eliminated in the current experimental setup. Furthermore, the uncertainty of detecting the filling level is difficult to determine but also affects the comparison of the experiments.

(II) **Physical effects as discussed in literature.** Theoretical and experimental analysis of the thin-film evaporation process suggests that a large share of the evaporation takes place at a rather small region in the vicinity of the 3-phase contact line where refrigerant vapour, liquid and the metal of the heat exchanger meet (cf. Section 2.2). In this region, the refrigerant film on the heat exchanger is comparably thin, yet sufficiently thick that intermolecular forces do not prevent evaporation ("disjoining pressure") (Xia et al., 2008; Crößmann, 2016; Fischer, 2015; Potash and Wayner, 1972). This region is affected by the curvature of the liquid–vapour interface and the apparent contact angle between the refrigerant and the heat exchanger. The contact angle, however, is not fixed as its hysteresis ranges from receding (minimum) over equilibrium to advancing (maximum) contact angle (Gao and McCarthy, 2007) and can additionally be affected by dynamics of the liquid (Eral et al., 2013) (cf. Section 2.2.2). In fact, in the experiments with continuously decreasing filling level, the refrigerant pool level is moving downwards. Thus, the 3-phase contact line that is formed between the tubes, the vapour and the refrigerant pool is also moving. Although the maximum speed of that specific 3-phase contact line is very slow with about $20\,\mathrm{mm\,h^{-1}}$, the movement itself might be sufficient to alter the contact angle and, thus, be partly responsible for the difference in measured U-values compared to the experiments with constant filling levels.

A difference in contact angle and, thus, capillary action, could also be the reason for the earlier detachment of the refrigerant from the tube in case of constant filling levels compared to continuously decreasing filling levels (Figure 4.5). However, the measured relative filling levels f_{rel} after detachment have to be considered with caution as the filling level will increase after detachment due to the volume of the detached meniscus that connected the refrigerant and the tube.

4.4 Conclusions

The first objective of this chapter was to identify and discuss potential impacts from experimental setups and procedures on the measurement of sub-atmospheric thin-film evaporation of water. To our knowledge, this was the first time that such a comparison has been conducted. Experimental results showed generally good agreement of the evaporation process. For lower driving forces, results from RWTH and ISE are identical within uncertainty of measurement for the logarithmic mean temperature differences, heat flows and U-values. Experiments with higher driving forces also showed generally good qualitative agreement of the evaluated evaporation quantities, yet revealed deviations in absolute U-value of up to 24 %. We believe that this discrepancy is most likely caused by different surface properties of the heat exchangers and remaining systematic differences between the measurement setups.

Based on our findings, we conclude that thin-film evaporation measurements can generally yield comparable results from different experimental setups. However, there are many parameters associated with the setup and procedure which are prone to be overlooked and which – if disregarded – can easily render a comparison of measurement results invalid. In particular, we consider the following aspects as very important:

The analysis of measurement uncertainties is crucial since small uncertainties can lead to larger uncertainties in calculated quantities due to propagation of uncertainties. In this work, the combination of high heat flows and small driving forces leads to both small temperature differences between in- and outlet temperature of the fluid and between fluid and saturation temperature. As a result, even the rather small uncertainties of the temperature sensors and pressure transducers can have a very large effect on the propagated uncertainties of the temperature differences.

For comparison of experiments, their input conditions need to be identical and if possible constant. This requirement holds for the evaporator inlet temperature, volume flow and evaporation temperature (saturation pressure in the evaporator).

For capillary-assisted thin-film evaporators, the surface properties of the investigated heat exchanger need to be exactly defined as surface properties can have a crucial effect on wetting behaviour, capillary action and consequently on evaporation performance. It should be taken into account that the manufacturing process, storage and measurement conditions can change the surface properties.

In order to contribute to the creation of correlations, it would be valuable to determine and provide contact angles for the investigated finned copper tubes at process conditions present during the experiments in future work. More information regarding contact angles can be found in Appendix D.5.

In general, control of non-condensable gases is a major challenge for evaporation experiments in sub-atmospheric pressure conditions. Non-condensable gases can distort the measurement of saturation pressure, impede condensation and possibly influence evaporation mechanisms. Therefore, vacuum tightness of the system and knowledge or control of leak rates are essential for reproducible experimental results.

The second objective of this work was to assess the impact of filling level dynamics: for the first time, we compared measurements at decreasing filling levels with experiments at constant filling levels. Results show that experiments with continuously decreasing filling level are well-suited for quickly characterising heat exchangers at all filling levels. However, for decreasing filling levels, measured U-values were always 5-10 % higher. Although the differences are close to or within uncertainty of measurement, excellent reproducibility of the experiments suggests that there are underlying systematic reasons. This effect could even be exploited in future thin-film evaporators: instead of evaporating at a constant filling level with continuous reflux, the evaporator could be operated with decreasing filling level and quick intermittent refills. Thereby, the maximum achievable U-value would increase by up to 10 %.

From our experience, we conclude that comparing absolute values obtained from different experimental setups requires a detailed look at experimental setup and conditions. Unfortunately, the required information is not always completely provided in literature. However, if all relevant influencing factors are considered in measurement procedure and evaluation, comparisons are possible.

In this chapter, the validation of the experimental setup and procedure for the evaporation experiments is successfully demonstrated: in general, reproducibility in repetitions of identical experiments in the setup at RWTH used in this thesis is excellent (cf. Figure 4.5). Furthermore, reproducibility is verified with another setup: experimental results from the setups at RWTH and ISE are identical within uncertainty of measurement for low driving force and still in good agreement for high driving force (cf. Figures 4.3 and 4.4). Finally, the analysis of filling level dynamics reveals that experiments with decreasing filling levels are well-suited for quickly characterising heat exchangers at all filling levels (cf. Figure 4.5). Thus, the experimental setup and procedure for evaporation experiments used in this thesis has been proven to generate reproducible results and, therefore, the validation was successful. In the next chapter, the validated experimental setup and procedure is employed to investigate capillary-assisted evaporation from coated tubes.

CHAPTER 5

Characterisation of capillary-assisted evaporation

This chapter addresses the first identified bottleneck of the refrigerant water for use in adsorption chillers: efficient sub-atmospheric evaporation of water. To overcome the challenges associated with sub-atmospheric evaporation of water (cf. Section 2.2.4), capillary-assisted evaporators maintain an evaporating thin film of water on their surface by capillary action. Required capillary action can be created by different structures as suggested in literature and introduced in Section 2.2.5:

(1) **Macroscopic structures**, e. g. finned tubes
(2) **Microscopic structures**, e. g. coatings
(3) **Combination of macroscopic and microscopic structures**, e. g. coated macroscopic structures

---

Contents of this chapter have in slightly amended form been reprinted from:

J. Seiler, F. Lanzerath, C. Jansen and A. Bardow. “Only a wet tube is a good tube: understanding capillary-assisted thin-film evaporation of water for adsorption chillers”. *Applied Thermal Engineering* 147 (2019), pp. 571–578, with permission from Elsevier. Contributions of the author: writing the draft as principal author, designing and building of the experimental setup, planning the measurement procedure, conducting the experiments, implementing the data reduction, and evaluating the results.

Reprinted passages use the *pluralis modestiae* “we” as an alternative to the passive voice.

Although all structures have been shown to improve the heat transfer for sub-atmospheric evaporation of water, coatings are especially promising since they can in principle be tailored to the application and applied to many different heat exchanger designs. However, commonly accepted design guidelines for coatings suitable for capillary-assisted, thin-film evaporation in adsorption chillers are lacking in the literature.

Besides heat exchanger structure, previous studies investigated the impact of many process parameters on capillary-assisted thin-film evaporation: (I) evaporation temperature, (II) driving force and (III) filling level. The summary of these studies in Section 2.2.5 revealed inconsistencies in the literature: effects regarding (II) driving force and (III) relative filling level are inconclusive in the published studies and, thus, require further investigation.

In this chapter, we investigate coated tubes and show that wetting of the tube is crucial for thin-film evaporation. Wetting of the tube is enabled by capillary action on the tube's coated surface. To analyse the impact of the microscopic structural properties of the coating on the wetting behaviour, we vary and experimentally examine the tube's coatings. We characterise the tubes' evaporation performance at different temperatures and driving forces for all filling levels. Our results show that dry-out of the heat exchanger surface can occur depending on driving force, temperature and surface structure, similarly to the capillary or boiling limit in heat pipes (Faghri, 1995) (cf. Section 2.2.3). The results allow us to resolve the conflicting findings in the literature discussed above. Finally, we correlate the tube surface properties with the measured evaporation performance of the tubes to provide a guideline for future thin-film evaporation evaporator design.

In Section 5.1, the used materials and methods for the experiments are described. Subsequently, the obtained results are presented and discussed in Section 5.2. The conclusions are summarised in Section 5.3.

5.1 Materials and methods

The experimental setup used for evaporation experiments is described in detail in Section 3.2. The experimental setup employed in this chapter differed slightly: the setup did not include the level drain (Figure 5.1). Hence, experiments with coated tubes at constant filling level are not presented. Nevertheless, the conclusions drawn from experiments with finned tubes (Chapter 4) should also hold for the coated tubes: experiments with continuously decreasing filling levels are well-suited for characterising heat exchangers at all filling levels as shown in Chapter 4. Thus, to quickly characterise the coatings at all filling levels, all experiments with coated tubes were conducted at decreasing filling levels.

The experimental setup was designed to determine the evaporation performance of the investigated tubes at different evaporator inlet temperature, driving force and filling level.

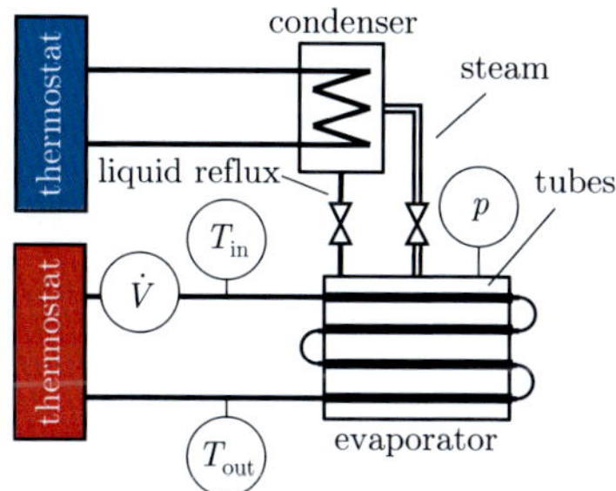

Figure 5.1: Experimental setup for measuring the overall heat transfer coefficient of the investigated tubes. Four identical tubes are serially connected and placed in a horizontal plane in the evaporator (figure shows top view of the evaporator). The condenser sets the driving force via its saturation pressure.

The evaporator was a vacuum-tight, cylindrical steel vessel that holds four identical tubes, which were serially connected. In this work, we study a single layer of tubes. In practice, several of such layers are usually stacked with a water pool for each layer. Thereby, compact designs can be achieved. Our analysis of a single layer therefore supports the improvement of practical evaporators. The investigated coated-copper tubes (inner diameter 13 mm, outer diameter $d_o = 15$ mm) were coated for a length L of 500 mm each. The connection of the tubes was located outside of the evaporator to enable easy replacement of the investigated tubes.

Thermal energy for the evaporation process was provided by water as heat transfer fluid flowing through the inside of the tubes. The water had a constant input temperature set by a thermostat.

5.1.1 Experimental procedure

The automated experimental procedure started by raising the filling level in the evaporator until the tubes were fully immersed in the refrigerant. It is important that the coatings are fully wetted prior to the start of the experiment, since the dry tubes are hydrophobic and not able to wet themselves by capillary action only. However, once the tubes have been immersed in the refrigerant and are fully wetted, they show hydrophilic behaviour.

The driving force of the evaporation process can be adjusted by the temperature difference between evaporator and condenser. With the steam valve still closed, the thermostats set the desired temperature in the evaporator and condenser. After the temperatures in the components were constant, the steam valve was opened and evaporation began. The water evaporated and the filling level decreased continuously, since the reflux valve was closed and no water was refilled into the evaporator. The experiment ended when the refrigerant lost contact to the tubes. This procedure allows to include all possible filling levels in one single experiment (Lanzerath et al., 2016). Technically, we did not measure steady-state values. However, all measured sensor values were in quasi-steady state condition. For more details, see Chapter 4.

5.1.2 Data reduction

The data reduction employed for the experiments is described in detail in Section 4.2.3. Here, only a short summary of the data reduction for the overall heat transfer coefficient U is given:

$$U = \frac{\dot{Q}}{A\,\Delta T_{\text{ln}}} = \frac{\dot{V}\rho_{\text{in}}\left[h_{\text{in}}(T_{\text{in}}) - h_{\text{out}}(T_{\text{out}})\right]}{n\pi d_{\text{o}}L \quad \frac{T_{\text{in}}-T_{\text{out}}}{\ln\left(\frac{T_{\text{in}}-T_{\text{sat}}(p)}{T_{\text{out}}-T_{\text{sat}}(p)}\right)}}. \tag{5.1}$$

with the measured quantities (all also shown in Figure 5.1) volume flow $\dot{V}$, in- and outlet temperature of the heat transfer fluid inside the coated tubes T_{in} and T_{out}, the pressure p in the evaporator and the required physical properties density ϱ, the specific enthalpy of the heat transfer fluid h and the saturation temperature of the refrigerant water T_{sat}. The reference surface area A was the total outer surface area calculated with the outer diameter d_{o} and the coated length L of each of the investigated $n = 4$ tubes.

5.1.3 Experiments

Experiments were conducted according to common operation conditions of evaporators in adsorption systems (Table 5.1). High volume flows for the inlet of the evaporator were chosen to achieve good heat transfer on the inside of the tubes. Each experiment was repeated at least three times. The standard deviations of the repeated experiments amplified by the corresponding value from the t-distribution from GUM (Joint Committee for Guides in Metrology, 2008) are mostly below the size of the symbols used in Figures 5.4 to 5.6. Therefore, we did not include error bars in the figures.

Table 5.1: Operating conditions of conducted experiments.

Evaporator inlet temperature T_{in}	**Set temperature difference between inlets of evaporator T_{in} and condenser $T_{in,cond}$** $T_{in} - T_{in,cond}$	**Evaporator inlet volume flow** $\dot{V}$
10 °C	2 K; 3 K; 4 K; 5 K	14.4 L min^{-1}
15 °C	2 K; 3 K; 4 K; 5 K	14.6 L min^{-1}
20 °C	2 K; 3 K; 4 K; 5 K; 10 K	14.9 L min^{-1}

We investigated 7 sets of tubes with coatings A–G and one set of plain uncoated tubes H. The coatings were characterised by microscopic metallographic analysis of cross-section polishes of the coated tubes to obtain surface properties. We selected properties which are influenced by the coating process and may have an impact on evaporation performance. Measured surface properties were surface extension, layer thickness, surface roughness R_z and porosity (Table 5.2 and detailed explanation in Table 2.1).

Table 5.2: Surface properties measured for tubes A–H in descending order of surface roughness. Detailed explanation of surface properties can be found in Table 2.1.

Tube	Unit	A	B	C	D	E	F	G	H
Surface roughness (R_z)	µm	72	71	61	56	47	38	37	5
Surface extension	-	1.71	1.53	1.69	1.53	1.48	1.47	1.27	1.08
Layer thickness	µm	218	282	105	189	49	89	322	0
Porosity	%	18.3	4.1	13.8	11.5	14.1	8.6	1.2	0
Symbol in Figures 5.5 and 5.6	-	×	△	□	◇	+	○	∗	▽

Surface extension is the ratio of the available heat transfer area and a perfect reference cylinder with the same diameter. The plain tube also has a surface extension > 1 indicating that the raw material for the coated tubes is not a perfect cylinder due to minor roughness of the raw material. Layer thickness of the coating is a key parameter of the coating process as it reflects the amount of coating material on the tube and, thus, also affects its production cost. Profile roughness and porosity are two important properties for the wetting and capillary action of the surface (Cho et al., 2016; Faghri, 1995). The profile roughness R_z used here is determined by averaging the maximum valley-to-peak distances of five consecutive sampling lengths of 100 µm. The porosity is the ratio of the available pore volume to the total volume in the coating determined by graphic analysis of the cross-section polishes.

5.2 Results and discussion

The conducted experiments vary in 4 dimensions: (1) evaporator inlet temperature T_{in}, (2) set driving temperature difference $T_{\text{in}} - T_{\text{in,cond}}$, (3) relative filling level f_{rel} and (4) tube coating A-H. In order to present the results of all conducted experiments, further data reduction is necessary.

To introduce the additionally employed data reduction, the impact of the driving force is shown exemplarily for the U-value of tube D for a constant evaporator inlet temperature in Section 5.2.1. The qualitative behaviour is similar for all other tubes. In a next step in Section 5.2.3, two characteristic values are identified for each experiment: the maximum UA-value $UA_{\text{eff,max}}$ and the corresponding filling level $f_{\text{rel}}(UA_{\text{eff,max}})$. These characteristic values are used below to compare different tubes.

5.2.1 U-value as function of driving force

The overall heat transfer coefficient U of tube D at 20 °C evaporator inlet temperature is shown in Figure 5.2 as a function of the relative filling level f_{rel}. The set temperature difference and, thus, also the driving force increases from dark (black) to lighter colours (light blue).

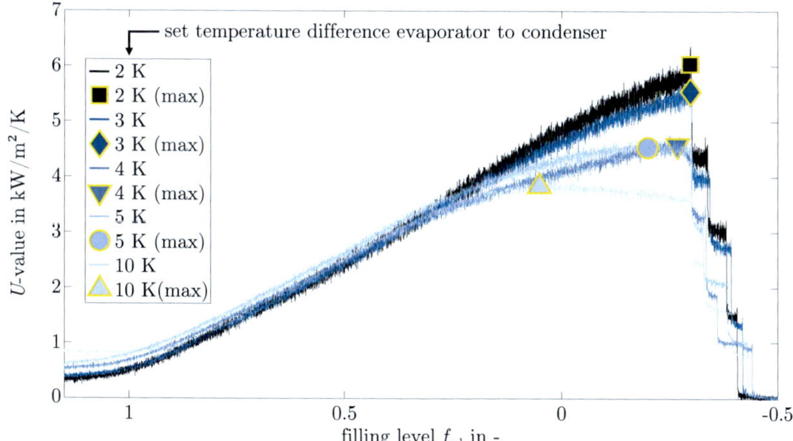

Figure 5.2: Overall heat transfer coefficient U for tube D at 20 °C evaporator inlet temperature and different set temperature differences (driving forces) as function of relative filling level f_{rel}. The set temperature difference increases from dark (black) to lighter colours (light blue). Maxima of U are marked with yellow-framed symbols and are used to characterise each experiment by two values: maximum U-value U_{max} and the corresponding filling level $f_{rel}(U_{max})$. The relative filling level f_{rel} is defined as function of outer tube diameter d_o in Figure 3.6 (c).

Generally, the U-value increases with decreasing filling level. At a negative relative filling level of about $f_{rel} = -0.3$, the U-value sharply drops in 4 distinctive steps: in each step, the contact between 1 of the 4 tubes and the water is lost. Once all tubes have lost contact to the water, they fall completely dry and the experiment is completed. At relative filling levels larger than 1, natural convection around the submerged tubes dominates the heat transfer as thin-film evaporation cannot take place. At high filling levels ($f_{rel} > 1$) experiments with higher driving forces (light blue) show larger U-values due to increased natural convection at higher temperature differences (Figure 5.2).

Decreasing the filling level creates a thin film as the tube emerges from the water. As the filling level decreases, the U-value increases, since thin-film evaporation offers higher U-values than natural convection. Lanzerath et al. (2016) showed that the increase in U-value with decreasing filling level is mainly due to the increase of available free surface area for thin-film heat transfer. At low driving forces, the U-value increases

wetted tube — partially dried out tube (same experiment)

Figure 5.3: Wetting of tube D at different filling levels: higher filling level and tube is fully wetted (left); lower filling level and tube is partially dried out on top (right).

until the tube's contact to the water is lost. However, at higher driving forces, maxima in U-value emerge (Figure 5.2). The maxima can be used to characterise each tube by two values: maximum U-value U_{max} and the corresponding filling level $f_{\mathrm{rel}}(U_{\mathrm{max}})$. With increasing driving force, the U_{max}-values decrease and occur at higher filling levels $f_{\mathrm{rel}}(U_{\mathrm{max}})$.

The reason for this effect is most likely a partial dry-out of the tube as can be seen in Figure 5.3 for tube D: at higher filling level, the tube is wetted as can be seen from light reflected from the tube's surface (Figure 5.3, left). At a lower filling level of the same experiment, the tube seems to be dry as the reflection of the light cannot be seen any more (Figure 5.3, right). Detailed analysis of the picture on the right-hand side shows that approximately the upper two-thirds of the tube have a different shading which suggests that this part is dried out.

At higher driving forces and, thus, also higher heat flows, capillary action on the tube's surface is not sufficient to keep the whole tube wetted. The mass flow of evaporating water on the tube exceeds the mass flow provided by capillary action. Consequently, the tube starts to dry out and thin-film evaporation does not take place in the dried-out areas of the tube. Thus, the total outer surface area A of the tube, which is used to calculate the U-value in Equation (5.1), is not the active surface area for heat transfer.

However, the effective surface area A_{eff} , at which thin-film evaporation takes place, cannot be determined in this experimental setup. Therefore, we consider in the following an effective, combined UA_{eff}-value instead of the U-value from Equation (5.1) to reflect the existence of possible dry-outs on the tubes:

$$UA_{\text{eff}} = \frac{\dot{Q}}{\Delta T_{\text{ln}}}. \tag{5.2}$$

5.2.2 Maximum UA_{eff}-values as function of driving force for different evaporator inlet temperatures

Again, each experiment can be characterised by its maximum UA_{eff}-value $UA_{\text{eff,max}}$ as well as the corresponding filling level $f_{\text{rel}}(UA_{\text{eff,max}})$ (cf. Figure 5.2). For tube A and B, these characteristic values are shown as function of the measured driving force ΔT_{ln} for different evaporator inlet temperatures in Figure 5.4.

For both tubes A and B, $UA_{\text{eff,max}}$ increases with higher evaporator inlet temperature T_{in}. This finding holds in fact for all tubes and is in agreement with other published studies. The increase in $UA_{\text{eff,max}}$ has most likely two main causes (Lanzerath et al., 2016; Thimmaiah et al., 2017; Xia et al., 2008):

(I) The heat transfer on the inner side of the tube is part of the overall heat transfer U and depends on temperature. With increasing evaporator inlet temperature T_{in}, the heat transfer on the inner side of the tube rises, leading to an increase in the overall heat transfer through the tube. A detailed description of this effect can be found in Lanzerath et al. (2016).

(II) Properties of water such as density and viscosity are also temperature-dependent and may affect the heat transfer into the thin film on the outside of the tube. The viscosity decreases with increasing temperature possibly leading to a lower resistance for the mass transfer through capillary action. The steam density generally increases with increasing temperature also resulting in a better heat transfer (Baehr and Stephan, 2011).

In contrast, the effect of driving force ΔT_{ln} differs for tubes A and B (Figure 5.4). While $UA_{\text{eff,max}}$ stays almost constant for tube A, it decreases with increasing ΔT_{ln} for tube B. The reason for the decrease of $UA_{\text{eff,max}}$ for tube B is a dry-out of the tubes as shown in Figure 5.3. Dry-out of tube A might occur at $\Delta T_{\text{ln}} > 3\,\text{K}$ but is hardly detectable in the conducted experiments.

Another indicator for dry-out is the filling level $f_{\text{rel}}(UA_{\text{eff,max}})$ at which $UA_{\text{eff,max}}$

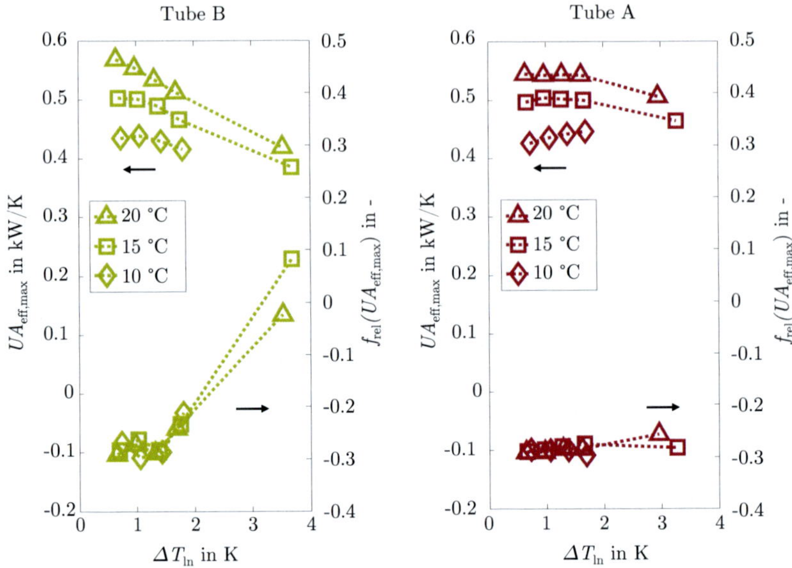

Figure 5.4: Maximum overall heat transfer coefficient multiplied with effective heat transfer area $UA_{eff,max}$ with corresponding relative filling levels $f_{rel}(UA_{eff,max})$ (axis on the right-hand side of the graphs) for evaporator inlet temperatures of 10, 15 and 20 °C as function of driving force ΔT_{ln} for coated tubes B (left) and A (right). Each experiment was repeated at least three times. Error bars are mostly below the symbols' size (cf. Section 5.1.3).

was detected: if $UA_{eff,max}$ does not occur at the lowest possible filling level, not all of the available heat transfer area on the outside of the tube is used and dry-out has most likely started before.

Although, theoretically, the available heat transfer area should not further increase below a relative filling level of $f_{rel} = 0$ (cf. Figure 3.6 (c)), the area occupied by the meniscus connecting the water to the tube continues to decrease for relative filling levels below zero. Thus, the available area for thin-film evaporation still increases even for relative filling levels below zero.

The measured relative filling levels $f_{rel}(UA_{eff,max})$ in Figure 5.4 show that $UA_{eff,max}$ starts decreasing at the same driving force ΔT_{ln} where $f_{rel}(UA_{eff,max})$ increases. Again, for tube A, a minor increase in $f_{\mathrm{rel}}(UA_{\mathrm{eff,max}})$ is present at $\Delta T_{\mathrm{ln}} > 3\,\mathrm{K}$. Nevertheless, tube A showed the best evaporation performance of all investigated tubes: good tubes, thus, work at high driving forces (heat flows) without dry-out. Therefore, in order to assess the performance of tubes, maximum UA_{eff}-values at high driving forces need to be evaluated.

The experiments conducted in this study suggest that the impact of the driving force ΔT_{ln} on the heat transfer coefficient in the investigated range ($\Delta T_{\mathrm{ln}} < 4\,\mathrm{K}$) is negligible as long as the thin film is present on the whole tube. In this case, the lowest possible filling level is optimal in terms of heat transfer. However, for reliable operation, it is important to ensure that the contact between the water and the tube is not lost and meniscus stability is maintained. Therefore, a higher filling level might be chosen as a trade-off between reliability and performance in practice.

Comparison of the findings with previous studies is difficult, since not all previous studies state whether dry-out occurred during the experiments. Xia et al. (2008), for example, measured lower U-values for higher driving forces for macroscopic structures. However, they did not mention whether dry-out occurred. Their findings are not in agreement with results from Lanzerath et al. (2016), who show that the driving force does not have an impact on the U-value. They observe the maximum U-value at the lowest possible filling level indicating that dry-out did not occur. Thimmaiah et al. (2017) determined the optimal relative filling level to be at 0.8. They report that the tube did not provide sufficient capillary action to fully wet the tubes at lower filling levels, thus parts of the tubes were dry in their experiments. Therefore, as mentioned before, the impact of the driving force is inconclusive in literature. Possible dry-out in the experiments by Xia et al. (2008) could, however, explain the different results: dry-out of the tubes could be responsible for the decrease in U-values for higher driving forces reported by Xia et al. (2008) as we measured the same effect for the investigated tubes in the current work.

Generally, previous studies agree that lower filling levels lead to higher U-values (Lanzerath et al., 2016; Xia et al., 2008), which is in agreement with our findings reported here. Our work now shows that this behaviour is found as long as the tube is fully wetted. If a thin water film is present on the tube, our results show that the UA-value for evaporation is independent of the driving force. The larger the available surface area for thin-film evaporation is, the larger the UA-value becomes. Therefore, it is important to create and maintain a thin film on the entire surface of the tube. The thin film is maintained by capillary action, which is governed by the characteristics of the surface. Therefore, we correlate the tubes' surface characteristics to the maximum UA-value in the next section.

5.2.3 Correlation between tubes' surface properties and maximum evaporation performance

For each tube, we correlate the evaporation performance measured by the $UA_{\text{eff,max}}$-value to the surface properties given in Table 5.2. Since measured $UA_{\text{eff,max}}$-values depend on evaporator inlet temperature T_{in} and driving force ΔT_{ln}, further data reduction is necessary to obtain one single evaluation parameter for each investigated tube.

As the influence of evaporator inlet temperature is similar for all tubes, we choose the medium temperature T_{in} of 15 °C for further data reduction. To merge the measurements with different driving forces ΔT_{ln} into one single value, we use a weighted average value. We select the logarithmic temperature differences ΔT_{ln} as weights, to reflect the difficulty to keep the tubes wetted at higher driving forces:

$$\overline{UA_{\text{eff,max}}} = \frac{\sum_i UA_{\text{eff,max}}(\Delta T_{\text{ln,i}})\Delta T_{\text{ln,i}}}{\sum_i \Delta T_{\text{ln,i}}}. \tag{5.3}$$

The calculated average maximum UA-values $\overline{UA_{\text{eff,max}}}$ are correlated to the 4 surface properties of the tubes A–H given in Table 5.2 (Figures 5.5 and 5.6). Tube A has the highest $(\overline{UA_{\text{eff,max}}})$-value of almost 0.5 kW K^{-1}. Tubes B, C, E and D have comparably high $(\overline{UA_{\text{eff,max}}})$-values, tubes F and G show comparably low $(\overline{UA_{\text{eff,max}}})$-values. The plain tube H with no coating to provide capillary action offers the lowest $(\overline{UA_{\text{eff,max}}})$-value of about 0.05 kW K^{-1}. Therefore, the coating improves the $(\overline{UA_{\text{eff,max}}})$-value by up to a factor 10.

Figure 5.5 demonstrates a positive correlation between surface roughness and the $(\overline{UA_{\text{eff,max}}})$-value and between surface extension and the $(\overline{UA_{\text{eff,max}}})$-value. Surface extension directly affects the available surface area for the heat transfer process. Thus, a positive correlation between surface extension and the $(\overline{UA_{\text{eff,max}}})$-value is probably causal. However, the increase in the $(\overline{UA_{\text{eff,max}}})$-value is larger than the proportional relationship of U multiplied by A suggests. Therefore, other effects should also be of significance. Surface roughness has been shown to amplify capillary forces (Chu et al., 2012). Therefore, the positive trend in the correlation between surface roughness and $(\overline{UA_{\text{eff,max}}})$-value might also be based on a causal connection.

Figure 5.6 (left) suggests that surface porosity also seems to be correlated to the $(\overline{UA_{\text{eff,max}}})$-value. In general, porosity might enhance mass transfer through the coating as porosity could create capillary action and increase the cross section available for mass transfer. Therefore, it seems reasonable that increased porosity could enhance the evaporation performance. However, closed porosity is not accessible for the refrigerant

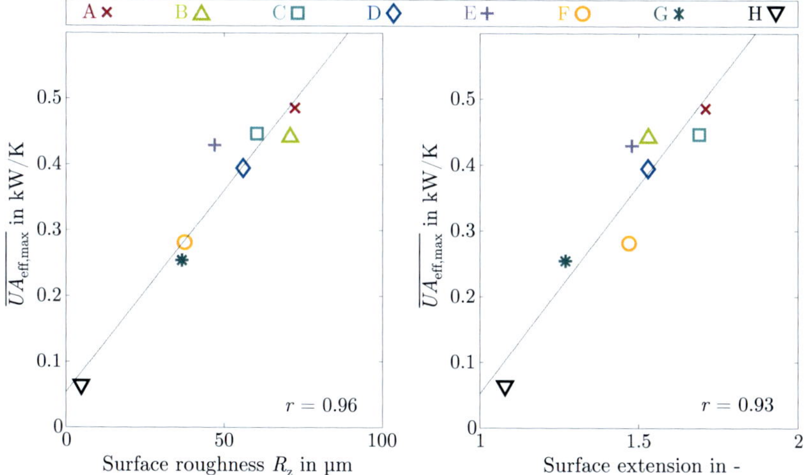

Figure 5.5: Average maximum overall heat transfer coefficient multiplied with effective heat transfer area $\overline{UA_{\mathrm{eff,max}}}$ as function of roughness R_z (left) and surface extension (right) for different tubes A–H. Linear correlation is given as black line with the Pearson correlation coefficient r.

and most likely increases thermal resistance. Thus, closed porosity should not be beneficial for evaporation performance and more experiments are necessary to deepen the understanding of porosity on the evaporation performance.

The layer thickness of the coating does not seem to be correlated to the ($\overline{UA_{\mathrm{eff,max}}}$)-value (Figure 5.6, right). Hence, we did not include a trend line in the figure. It could be possible that lower layers of the coating are not accessible for the refrigerant with the given surface properties, e. g. porosity. Therefore, thin coatings are favourable, as layer thickness does not have a significant impact on performance and costs as well as thermal resistance can generally be reduced with thinner coatings.

It is to be noted that the best evaporation performance was measured at the highest value of all three correlated surface characteristics: surface roughness, extension and porosity. Therefore, it seems likely that further improvements are possible which should be experimentally evaluated.

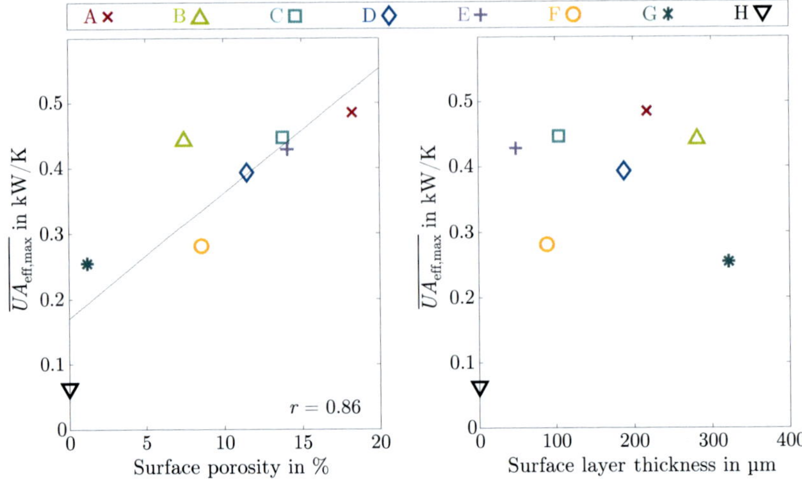

Figure 5.6: Average maximum overall heat transfer coefficient multiplied with effective heat transfer area $\overline{UA_{\mathrm{eff,max}}}$ as function of porosity (left) and layer thickness (right) for different tubes A–H. Linear correlation is given as black line for porosity with the Pearson correlation coefficient r (left). No reasonable linear correlation is found for layer thickness (right).

As mentioned before, it is important to create and maintain a thin film on the entire surface of the tubes to achieve good evaporation performance. In this work, the thin film is solely maintained by capillary action on the tubes. However, we also tested the evaporation on coated tubes in the evaporator installed in a real lab-size adsorption heat pump and observed an interesting phenomenon: the refill process of water from the condenser to the evaporator can be designed to create a splash of water that wets the surface of the evaporator tubes. The coating on the tubes then distributes the water via capillary action creating a thin film on the tubes' surface. Thus, the capillary action of the coating is supported by the splash of the refill process. This effect could be incorporated in future thin-film evaporation design.

5.3 Conclusions

Thin-film evaporation has been shown to allow for efficient evaporation of water in adsorption chillers. To improve the understanding of the underlying phenomena, we experimentally investigated thin-film evaporation from coated copper tubes. The evaporation performance of 7 coatings and a plain tube were determined for all filling levels at different temperatures and driving forces. Evaporation performance was then correlated with surface characteristics of the tubes. The conducted experiments contribute the following key results:

(I) The investigated coatings increase the UA-value compared to plain tubes by a factor of up to 10.

(II) Increasing evaporator inlet temperatures lead to higher UA-values.

(III) The optimal filling level should be as low as possible to exploit the maximal available heat transfer area for thin-film evaporation. For reliable operation, however, higher filling levels can be reasonable to ensure continuous contact between tube and refrigerant (meniscus stability).

(IV) The impact of the driving force on the U-value is not detectable as long as the tubes are kept fully wetted by sufficient capillary action. At high driving forces, dry-out of the tubes decreases the U value.

(V) Good coatings are able to maintain a thin film on the tubes' surfaces by capillary action even at higher driving forces. The investigated coatings vary considerably in their ability to maintain a thin film.

(VI) Higher porosity, surface extension and roughness improve the evaporation performance as shown by positive correlations between porosity, roughness, and surface extension and evaporation performance. Coating layer thickness can be thin as no impact on evaporation performance was found.

The investigated thin-film evaporation increases the overall heat transfer coefficient for sub-atmospheric evaporation about one order of magnitude compared to pool evaporation with a plain tube. Hence, evaporators with higher power densities are rendered possible to improve the performance of adsorption heat pumps. To tap this potential, the creation of a thin film on the heat exchanger is crucial. Our results show: only a wet tube is a good tube!

In the conducted experiments, coatings proved to be a versatile option to create capillary action on a simple heat exchanger for exploiting high UA-values by thin-film evaporation. Evaporation performance of the investigated coatings depends strongly on driving force: at low driving forces, most coatings were fully wetted; however, at high driving forces, sub-optimal coatings suffered from dry-out and only good coatings continued to keep the tubes' surfaces wetted. High driving forces also occur in the evaporator of an adsorption chiller, as its cyclic operation also leads to high driving forces at the beginning of the adsorption process. Thus, thin-film evaporation performance of heat exchangers designed for adsorption chillers should always be assessed at varying conditions including high driving forces.

To further enhance the evaporation performance of the investigated coatings, deeper understanding of the underlying evaporation mechanisms present in the conducted experiments would be valuable. Although the experiments were not designed to investigate evaporation mechanisms, analysis of the photos of the conducted experiments gives some insight.

The photos in Figure 5.3 show menisci between the refrigerant pool and the coated tubes. These menisci probably contributed to thin-film evaporation, which often occurs in the context of extended menisci (cf. Section 2.2.6). However, evaporation from these menisci was probably not sufficient to explain the strongly increasing UA-value for lower filling levels: the shape of these menisci only slightly changed with changes in filling level and, especially at low filling levels, thermal connection between the relatively large menisci and the tube is most likely limiting. Instead, the governing effect for the strongly increasing UA-value is most likely connected to the increase of available surface area for lower filling levels as described above. Hence, it seems as if thin-film evaporation on the tubes' surface mainly contributed to evaporation.

In the context of thin-film evaporation from an extended meniscus in nucleate boiling, two possible evaporation mechanisms have been discussed in literature: contact line evaporation vs. microlayer evaporation. Photos on which the coated tubes macroscopically seem to be fully covered with reflective water (cf. Figure 5.3 (left)), could suggest that evaporation took place in a microlayer. However, it is also impossible to exclude the existence of contact lines as the quality of the pictures is not sufficient to optically resolve possible contact lines.

Capillary action is the only mechanism known to explain the continued wetting of the tubes. Consequently, capillaries need to be present on the coated surfaces and evaporation probably also occurs in extended menisci of these capillaries. Shape, size and length of the menisci involved in evaporation, however, could not be determined with the measurement equipment accessible for the work in this thesis. Instead, integral surface quantities were evaluated and correlated to evaporation performance to give practical advice on future evaporator development: coatings should be tailored to increase surface roughness, extension and porosity. These properties probably promote the creation of capillaries and menisci responsible for high UA-values of thin-film evaporation. Thicker layers and closed porosity, however, are probably not beneficial for evaporation performance, since they increase thermal resistance.

Chapter 6

Water-based adsorption cooling below 0 °C

In this chapter, the second identified bottleneck of the natural refrigerant water in adsorption chiller is addressed: limitation to temperatures above 0 °C due to freezing of the refrigerant.

First, the state-of-the-art regarding low temperatures for water-based adsorption chiller is summarised and an overview regarding alternative working pairs is given. Next, to overcome the limitation of the refrigerant water, the solution proposed in this thesis is introduced and theoretically discussed: adding the anti-freezing agent ethylene glycol to the evaporator to prevent freezing. To realistically assess the potential for low temperature operation, experiments with a water-based adsorption chiller were performed and the results are presented and evaluated.

Contents of this chapter (Section 6.1 to Section 6.5) have in slightly amended form been reprinted from:

J. Seiler, J. Hackmann, F. Lanzerath, and A. Bardow. "Refrigeration below zero °C: Adsorption chillers using water with ethylene glycol as antifreeze". *International Journal of Refrigeration* 77 (2017), pp. 39–47, with permission from Elsevier. Contributions of the author: writing the draft as principal author, adapting the design and building of the experimental setup, planning the measurement procedure, supervising the experiments, implementing the data reduction, and evaluating the results.

Reprinted passages use the *pluralis modestiae* "we" as an alternative to the passive voice.

6.1 Introduction

Preferably, the natural refrigerant water is used as working fluid in adsorption chillers since it is environmentally safe and offers high enthalpy of vaporisation (Fernandes et al., 2014). However, many applications require evaporator temperatures close to or below the freezing point of water. Freezing of the refrigerant, though, causes a breakdown of the adsorption chiller output. Thus, using pure water as working fluid is not feasible in adsorption processes close to 0 °C due to the location of the triple point of water (Freni et al., 2015; Wang et al., 2009, 2014b).

So far, the only solution for adsorption applications at temperatures below 0 °C is to use a different refrigerant than water, e. g. methanol (Restuccia et al., 2005) or ammonia (Miles and Shelton, 1996). However, switching refrigerants for (occasional) low-temperature operation requires specific adsorption chiller designs and, thus, affects the whole setup. Li et al. (2014) give an overview of solid–gas sorption working pairs pointing out certain advantages and drawbacks of the common low-temperature refrigerants: ammonia is toxic whereas methanol is flammable and restricted to temperatures below 120 °C due to decomposition (Hu, 1998). In contrast, ammonia is stable at high temperatures and, thus, it is also suitable for high temperature heat sources. Ammonia offers the advantage of comparably high saturation pressures which allows for simpler constructions compared to vacuum-tight constructions for methanol and water. However, ammonia has a pungent odour and is incompatible with copper (Li et al., 2014; Wang et al., 2014b). Furthermore, methanol and ammonia only offer a low enthalpy of vaporisation compared to water (Jiao, 2013). To compensate for the lower enthalpy of vaporisation, adsorbents with higher refrigerant uptakes have been investigated, e. g. metal organic frameworks (Rezk et al., 2013). However, to clearly outperform state-of-the-art adsorbents, metal organic frameworks need to be tailored to the application and require increased heat and mass transfer as well as increased densities (Graf et al., 2020; Liu et al., 2020).

In this work, we propose an alternative strategy for adsorption chillers at 0 °C: we suggest using water mixed with an anti-freezing agent to lower the freezing point of the refrigerant. Using a mixture of anti-freezing agent and water aims at retaining the good adsorption properties of the refrigerant water in the adsorption cycle at temperatures above the freezing point while resolving the issue of freezing in the evaporator for temperatures below 0 °C.

In this study, we use an ethylene glycol–water mixture to expand the operational range of a zeolite adsorption chiller below 0 °C. We show that the adsorption chiller process still works with ethylene glycol (in the following referred to as glycol) as anti-freezing agent in the evaporator. Furthermore, we show that the glycol practically remains in the evaporator and, therefore, water is used as refrigerant in the adsorption chiller cycle itself. Although results are presented for an adsorption chiller, they are also relevant for adsorption heat pump applications with (occasional) sub 0 °C ambient temperature heat sources. Experiments were conducted using our lab-scale adsorption chiller setup.

In Section 6.2 and Section 6.3, the properties of the working fluid mixture used as well as the experimental setup are described, respectively. In Section 6.4, experimental results are presented and discussed for different glycol–water mass fractions. Conclusions are drawn in Section 6.5.

6.2 Working fluid mixture glycol–water

Addition of glycol to water lowers the freezing point of the mixture which, thus, stays in the liquid phase at temperatures below 0 °C. The solid–liquid phase diagram of the mixture glycol–water is shown in Figure 6.1 (a). Due to the existence of two eutectic points, the freezing points of mixtures with more than 30 mass-% glycol lie below the freezing points of both pure components. This effect is the reason that mixtures of glycol water are well suited for anti-freezing applications (Curme and Young, 1925). Therefore, the proposed addition of glycol into the evaporator should reliably prevent freezing of the refrigerant in adsorption chillers.

Adsorption of glycol on zeolites has been studied before: Erdem et al. (2006), Erdem and Michel (2005) and Sharma et al. (2010) analysed the adsorption of glycol on zeolites NaX and H-ZSM5 experimentally to understand and model the molecular interaction between glycol and zeolite. Their experiments show that glycol can be adsorbed by zeolites. Hence, the question remains whether the zeolite NaY employed here will also adsorb glycol from a glycol–water mixture. To estimate the potential influence of glycol on the adsorption process, the vapour phase of the glycol–water mixture is examined. Experimental data from Sokolov et al. (1971) shown in Figure 6.1 (b) indicates that, for liquid glycol fractions below 60 mass-%, the vapour phase mostly consists of water, while glycol stays in the liquid phase: assuming an ideal vapour phase the molar composition is directly proportional to the partial pressure. Hence, the partial pressure of glycol of the vapour mixture should be comparably low. Therefore, it is unlikely that large amounts of glycol will be adsorbed which makes glycol a promising anti-freezing agent.

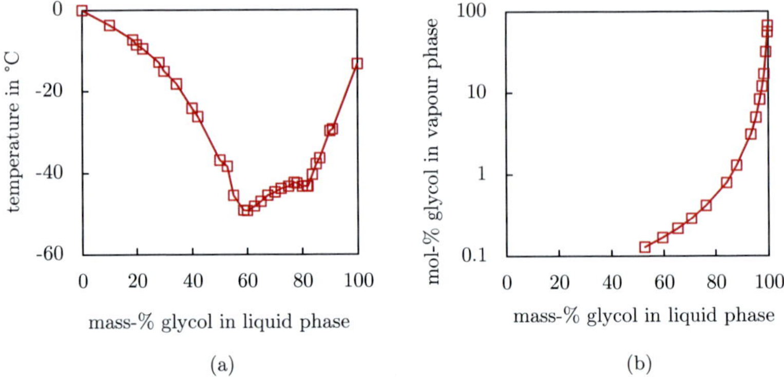

Figure 6.1: (a) Solid–liquid phase equilibrium data for the mixture glycol–water with experimental data from Cordray et al. (1996). (b) Mole fraction of glycol in vapour phase for different glycol mass fractions in liquid phase for an isobar at 1.3332 kPa with experimental data from Sokolov et al. (1971).

However, the addition of glycol to the evaporator does not only affect the composition of the vapour phase but also reduces the total pressure. Experimental data for isothermal vapour pressure of the glycol–water mixture (Horstmann et al., 2004; Nath and Bender, 1983; Villamanan et al., 1984) shows almost no deviation from the ideal Raoult's law (Çengel and Boles, 2010). Thus, we use Raoult's law to estimate the total pressure for the glycol–water mixture at 5 °C with the vapour pressure of the pure components (Lindstrom and Mallard, 2013; pro KÜHLSOLE GmbH, 2012). The evaporation temperature of 5 °C is the lowest reasonable temperature at which we can compare pure water as refrigerant to a glycol–water mixture as refrigerant: if we run the adsorption process at a lower evaporation temperature with pure water, there is the risk of freezing in the evaporator. We show the impact of glycol addition for the highest glycol fraction of 60 mass-% leading to the largest possible drop in vapour pressure. As can be seen in Figure 6.2, the drop in vapour pressure Δp due to the addition of 60 mass-% glycol to pure water is about 2.6 mbar.

The partial pressure of water is crucial for the adsorption cycle as it determines the driving force and the adsorption equilibrium state of the adsorption cycle. To quantify the impact of lower partial water pressure on the adsorption equilibria, an idealized equilibrium adsorption cycle is examined: the lowered vapour pressure for 60 mass-% glycol–water mixture from Figure 6.2 is used as the partial water vapour pressure in the

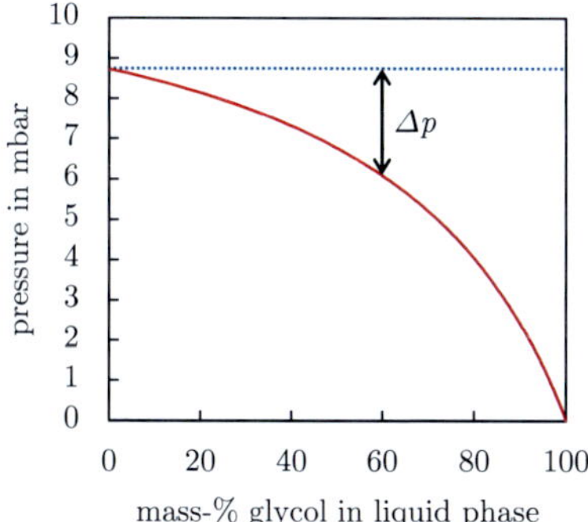

Figure 6.2: Ideal vapour pressure of glycol–water mixture for different glycol mass fractions (solid red line) and pure water (dotted blue line) at 5 °C calculated with Raoult's law from the pure components vapour pressure (Lindstrom and Mallard, 2013; pro KÜHLSOLE GmbH, 2012).

evaporator. Pure water adsorption on zeolite NaY is modelled with equilibrium states from experimental data (Schawe, 2001), assuming that glycol does not participate in the adsorption process. Two idealized equilibrium adsorption cycles with water vapour pressures for pure water (dotted blue line) and 60 mass-% glycol–water mixture (solid red line) at 5 °C in the evaporator are shown in an isosteric diagram in Figure 6.3.

The difference in water vapour pressure in the evaporator causes a difference in the maximum loading Δw of the two processes of 0.006 kg kg^{-1}, which corresponds to a decrease of 5.6 % in total loading difference between adsorption and desorption equilibrium loading. For lower desorption temperatures, this percentage increases. Lower total loading difference directly translates into less adsorbed refrigerant and, thus, leads to less cooling power provided. The expected decrease for equilibrium cycles of about 6 % is comparatively low and experiments are necessary to verify this estimate as well as the impact of lower water vapour pressure on the kinetics of real cycles. Therefore, we introduce the experimental setup employed in this work in the next section.

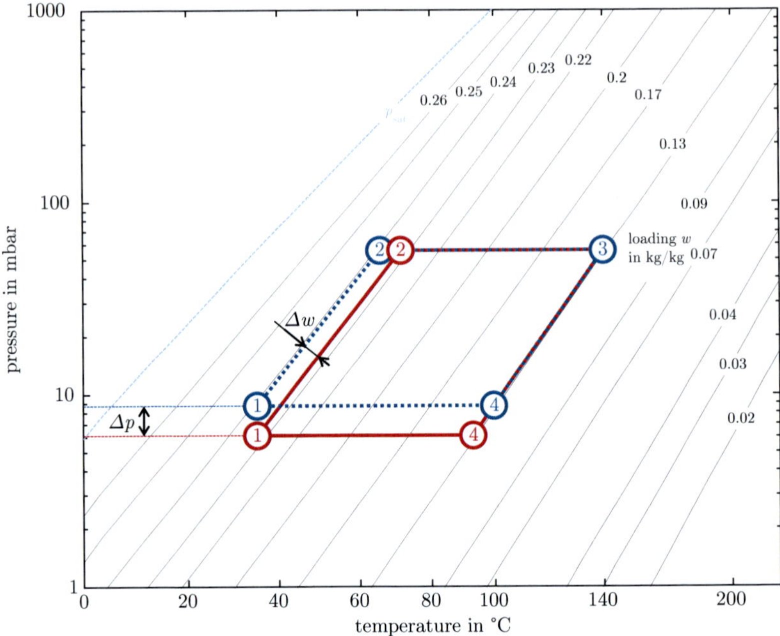

Figure 6.3: Isosteric diagram ($\ln p - (-1/T)$) for ideal NaY–water and NaY–(glycol–water) equilibrium adsorption cycle (experimental data from Schawe (2001) with (solid red line) and without glycol (dotted blue line) in the evaporator assuming that glycol does not participate in adsorption process. Black lines are isosters, i. e. states with the same loading w, and the saturation pressure of water is the dashed light blue line. Evaporation temperature is 5 °C.

6.3 Experimental setup

Experiments were carried out with a lab-size one-bed adsorption chiller (cf. Section 3.1.1). The chiller consists of the components evaporator, adsorber and condenser, which are connected by controlled steam valves (Figure 6.4). All components are thermally insulated to minimise heat losses to and from the environment.

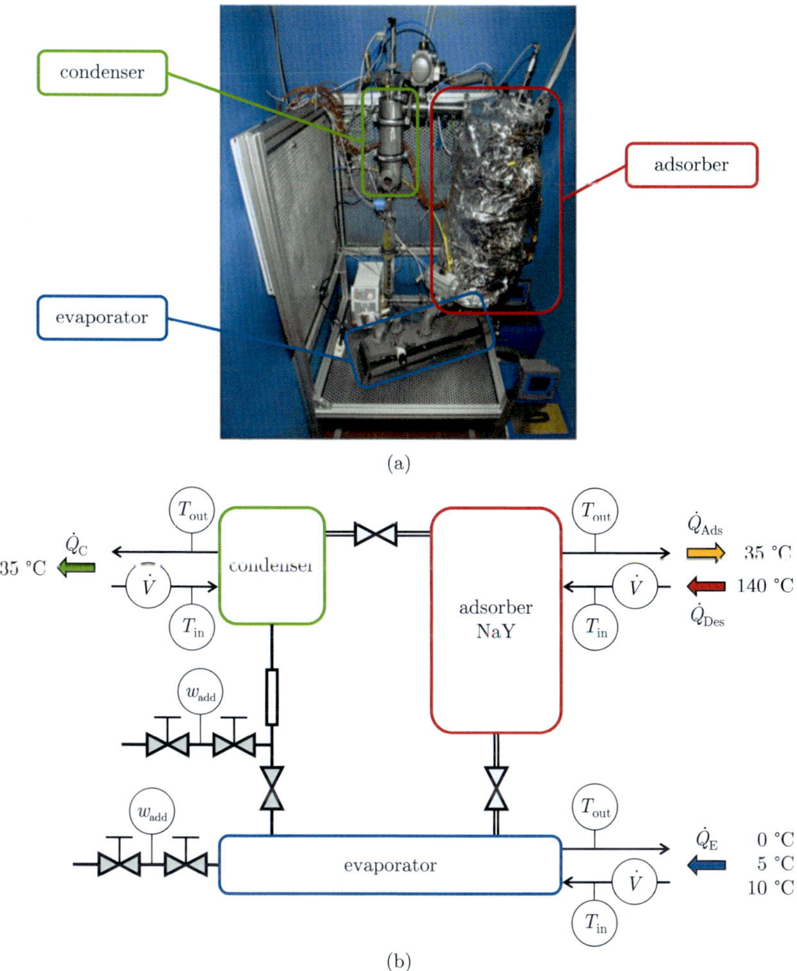

Figure 6.4: Photo of the experimental setup (a) and layout of the adsorption chiller setup and investigated temperature levels (b).

The experiments are conducted according to the following adsorption cycle:

During the adsorption phase, the working fluid is evaporated. The evaporated water steam flows through the opened valve into the adsorber. There, the water is adsorbed and the adsorption enthalpy is released. The adsorber contains about 5 kg zeolite NaY on a lamellae heat exchanger, which is cooled during the adsorption phase. The desorption phase is started by heating the adsorber. As soon as the pressure in the adsorber exceeds the condenser pressure, the valve to the condenser is opened. Hence, the steam valves are operated as if they were flaps reacting to a difference in pressure. Released water vapour streams from the adsorber into the condenser where the water is condensed. A measuring cylinder collects the liquid reflux from the condenser and allows taking samples of the working fluid. The working fluid is pumped back from the measuring cylinder into the evaporator before the next adsorption phase. Thus, a defined filling level of the evaporator is maintained at the beginning of each adsorption phase.

The experimental setup including employed sensors and data reduction is described in detail in Section 3.1. To evaluate the performance, heat flows $\dot{Q}_{\text{component}}$ and the transferred heats $Q_{\text{component}}$ during the cycles are determined for each component of the adsorption chiller. The cycles are taken from a series of cycles once cyclic steady-state has been reached.

Measurement uncertainties for the transferred heats are determined using the uncertainty of the calibrated sensors. The calculations have been carried out according to "Guide to the expression of uncertainty in measurement" (GUM) using a coverage factor of $k = 1$ (Joint Committee for Guides in Metrology, 2009; Lanzerath, 2014) (cf. Section 3.3). The uncertainties of the measurements are presented as error bars in the diagrams. They are typically below 10 %, however, become larger for lower heat flows.

Samples of glycol–water mixtures were taken from the evaporator and the measuring cylinder to determine the glycol–water mass fraction. The measurements were conducted using a density measuring module (Anton Paar, type DMA 4500 M).

The experimentally investigated temperature levels are shown in Figure 6.4 (b). Desorption and adsorption phase lengths were set to 700 s to represent a medium cycle time for the given size of the adsorber. The phase time was kept constant.

Experiments were carried out using glycol–water mixtures with 0, 30 and 60 mass-% glycol in order to evaluate the influence of the glycol mass fraction on the process. According to the phase equilibrium diagram in Figure 6.1 (a), the glycol–water mass fraction of 30 % has a freezing temperature of approximately −12 °C while the lowest possible freezing temperature of −50 °C is reached at a glycol mass fraction of 60 %.

Table 6.1: Operating conditions of conducted experiments (cf. Figure 6.4).

	Evaporator	Adsorption	Desorption	Condenser
Inlet temperature	0, 5 and 10 °C	35 °C	140 °C	35 °C
Glycol mass fraction	0 %, 30 % and 60 %	-	-	-

6.4 Results

The setup and working fluid described in Sections 6.2 and 6.3 were used to carry out experiments using the specifications given in Table 6.1 and Figure 6.4 (b).

The experimental results show that glycol can be used as anti-freezing agent in the adsorption chiller using zeolite–water as working pair. Furthermore, the influence of glycol mass fraction on the adsorption chiller process is studied in Section 6.4.1. As adsorption chillers may be operated under varying conditions, results for non-freezing operational conditions are also presented and discussed in Section 6.4.2.

6.4.1 Glycol as anti freezing agent

As described in Section 6.2, the use of pure water is infeasible in adsorption chiller processes at evaporating temperatures around 0 °C due to freezing of the refrigerant. Addition of the anti-freezing agent glycol to the evaporator, however, overcomes this limitation as Figure 6.5 and Figure E.1 in Appendix E demonstrate. For the shown experiment, the temperatures of both the liquid refrigerant in the evaporator and the evaporator outlet are below 0 °C.

Successful operation of the adsorption process is shown by the heat flows of the evaporator/condenser for the adsorption/desorption phase in Figure 6.5. Thus, using glycol as an anti-freezing agent renders evaporating temperatures below 0 °C possible.

In Figure 6.5 (a), the reflux from the measuring cylinder (condenser) into the evaporator causes the downward peaks between 800 s and 1000 s and lowers the transferred heats. Within the shown uncertainty, the measured heats during evaporation and condensation are in agreement with the enthalpy of vaporisation/condensation of water at the given temperatures. All measurements show slightly lower transferred heats in the evaporator than in the condenser, although the enthalpy of vaporisation/condensation for water increases with lower temperatures. Therefore, a systematic difference of the

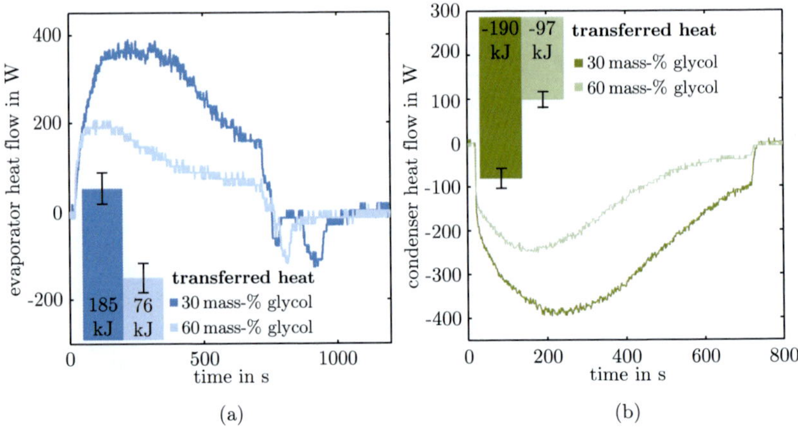

Figure 6.5: Heat flows and transferred heats at 0 °C in evaporator (during adsorption phase, (a)) and condenser (during desorption phase, (b)) for 30 % and 60 % glycol mass fraction. Error bars specify the uncertainty of measurement.

heat flow measurement in the two components is present. Presumably, the difference is caused by the vortex flowmeter in the evaporator in combination with the heat transfer fluid Glykosol N. However, this error does not affect the conclusions regarding the comparisons of different experiments.

Experimental settings for the two experiments shown in Figure 6.5 are identical except for the mass fraction of glycol. In case of a higher glycol mass fraction, heat flows are lower in the evaporator and condenser. Thus, less water is evaporated and adsorbed in the adsorption phase. Consequently, during the desorption phase, less water is condensed in return and heat flows in the condenser are also reduced. Therefore, the overall power is reduced.

Additionally, samples from the evaporator and the condenser reflux have been analysed with respect to composition (Table 6.2). Compositions are similar in the evaporator before and after the experiment.

The slight increase in glycol mass fraction in the evaporator for the 30 mass-% experiment to 33.8 mass-% is most likely caused by the uncertainty of the filling level measurement: the filling level of the evaporator can vary slightly. Thus, less water can be present in the evaporator which leads to a higher glycol mass fraction. Both before and after the experiments, the reflux from the condenser is pure water (<0.1 mass-%

Table 6.2: Measured glycol mass fractions in evaporator and condenser before and after conducted experiments (>40 cycles each) with 30 and 60 % glycol mass fraction in the evaporator. Lower detection limit is 0.1 mass-%.

Experiment	Before experiments		After experiments	
	Evaporator	Condenser	Evaporator	Condenser
30 mass-%	29.0 mass-%	<limit	33.8 mass-%	<limit
60 mass-%	60.0 mass-%	<limit	59.6 mass-%	<limit

glycol is the lower detection limit of the instrument) indicating that glycol has neither been adsorbed nor desorbed. The mass fraction measurements support our assumption that water is used as a refrigerant while glycol remains in the evaporator.

6.4.2 Performance for non-freezing operational conditions

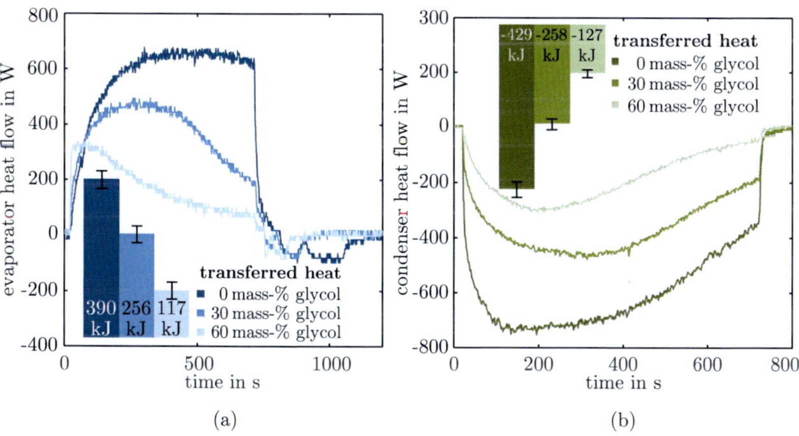

Figure 6.6: Heat flows and transferred heats at 5 °C in evaporator (during adsorption phase, (a)) and condenser (during desorption phase, (b)) for 0 %, 30 % and 60 % glycol mass fraction. Error bars specify the uncertainty of measurement.

For non-freezing operation conditions, an adsorption chiller using a glycol–water mixture in the evaporator will compete with pure water as refrigerant. Therefore, we compare 30 mass-% and 60 mass-% glycol–water mixtures with pure water as refrigerant at an evaporation temperature of 5 °C. Experiments are carried out using the same experimental conditions as before. The heat flows and transferred heats are shown in Figure 6.6.

The glycol mass fraction influences the heat flows in the same way as observed from the 0 °C experiments: increasing glycol mass fractions also reduce heat flows at 5 °C. Using a 30 mass-% glycol–water mixture decreases the transferred heat by 34 % compared to the experiment with pure water; a 60 mass-% mixture leads to a 70 %-decrease of transferred heat in the evaporator compared to pure water. The reductions in transferred heats in the evaporator and in the condenser correspond well to each other and therefore demonstrate consistency of the measured data.

Increasing glycol mass fractions could in principle reduce the transferred heats by two mechanisms:

(I) **Changed kinetics of the process.** The decreased transferred heats could result from reduced heat transfer properties of the glycol–water mixture. In general, heat transfer coefficients can be considerably smaller for mixtures (Baehr and Stephan, 2011): addition of glycol decreases the heat transfer coefficient of the mixture and, thus, hinders the heat transfer which then results in lower heat flows. Furthermore, the pressure difference between the adsorber and the evaporator during adsorption is lower due to the vapour pressure reduction caused by the addition of glycol as shown in Section 6.2. Thus, the driving force for the evaporation of water is decreased as well, also resulting in lower heat flows.

(II) **Influence of glycol on the water–zeolite thermodynamic adsorption equilibria.** The vapour pressure reduction caused by the addition of glycol has a low impact on the thermodynamic adsorption equilibria as shown in Section 6.2. However, this estimation is subject to the assumption that glycol does not participate in the adsorption process. Despite the low presence of glycol in the vapour phase of the evaporator, glycol could still be adsorbed by the zeolite. This co-adsorption of water and glycol might change the equilibria which would then affect the measured transferred heats.

To analyse the influence of the glycol–water mixture on the thermodynamic equilibrium, additional experiments have been conducted. For these experiments, the phase times were increased until heat flows in all components were negligible and the process was near thermodynamic equilibrium. The reflux from the condenser into the evaporator was active throughout the adsorption process to maintain a constant filling level in the evaporator. These experiments were carried out for pure water and a glycol–water mass fraction of 30 % at an evaporator temperature of 10 °C. Heat flows and transferred heats of the evaporator and condenser for the equilibrium experiments are given in Figure 6.7.

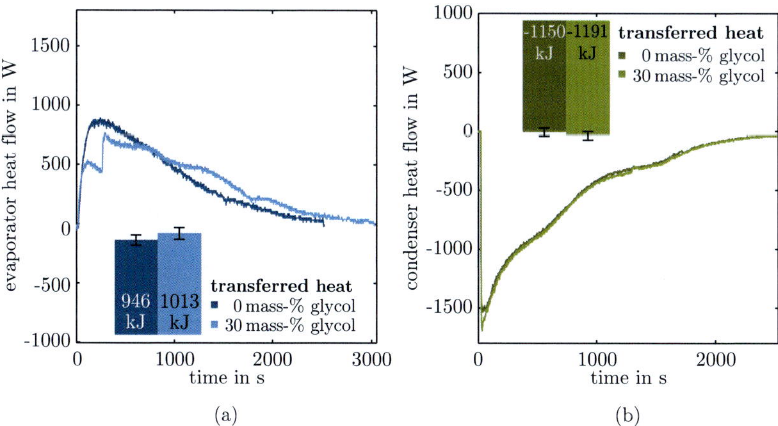

Figure 6.7: Heat flows and transferred heats at 10 °C in evaporator (during adsorption phase, (a)) and condenser (during desorption phase, (b)) for 0 % and 30 % glycol mass fraction for cycle times reaching equilibrium. Error bars specify the uncertainty of measurement.

The heat flows for evaporator and condenser in Figure 6.7 differ from the heat flows in Figure 6.5 and Figure 6.6: As the phase times are increased, the adsorber exhibits a larger loading difference. The larger loading difference leads to higher driving forces at the beginning of the phases and, thus, the peaks of the heat flows are more pronounced.

For the condenser, Figure 6.7 (b), measured heat flows are almost identical for pure water and 30 % glycol mass fraction. For the evaporator, Figure 6.7 (a), the heat flows differ substantially between pure water and 30 % glycol mass fraction, though, the transferred heats are almost the same.

An estimation given by the ideal adsorption cycle according to the calculation shown in Section 6.2 predicts a decrease in transferred heats of about 1.8 %. Since this change is within the measurement uncertainty of our experimental setup, it is too small to be detected. Thus, we cannot measure any influence of glycol on the adsorption equilibria. Consequently, our assumption is supported that glycol does not participate in the adsorption process, and we can assume that glycol has a low impact on the adsorption equilibria. As Freni et al. (2015) show, the maximum COP for a given system in an ideal cycle only depends on the adsorption equilibria. Therefore, the addition of glycol should also have a low impact on the maximum COP.

The kinetics of the evaporation process, however, are most likely influenced by the glycol. Change in the dynamic behaviour of the evaporator is possibly triggered by kinetics of the heat transfer: the rapid change in the heat flow around 400 s in Figure 6.7 (a) is reproduced in every experiment and probably due to a rapid change of the effective heat transfer area. This change might be due to locally different glycol–water mass fractions caused by the reflux which modifies the capillary action of the glycol–water mixture in the axially-finned heat exchanger. Since only water evaporates on the surface, the concentration of glycol increases locally. The resulting concentration polarisation also represents a further resistance for the evaporation process since the water needs to be transported to the surface (cf. Section 7.2).

We conclude that the main cause for the performance decline with increasing glycol mass fraction is a change in the kinetics of the evaporation/adsorption process due to the presence of glycol in the evaporator. The change in the kinetics can be attributed to two main side effects caused by the addition of glycol:

(I) Loss of driving force for the evaporation process due to reduced vapour pressure

(II) Lower heat transfer coefficients of the glycol–water mixture

(I) can only be addressed by modification of the adsorbent–adsorbate pair, whereas (II) can be compensated by increasing the heat/mass transfer in the evaporator. The improvement could be realised, for example, by using a larger heat transfer area or different evaporator heat exchanger geometry. Thereby, the anti-freezing agent glycol could be used while retaining the performance of water-based adsorption chillers.

6.5 Conclusions

The limit of water-based adsorption chillers to work around 0 °C evaporation temperature is overcome by using glycol as anti-freezing agent in the evaporator. Experimental results indicate that glycol remains in the evaporator and water is used as refrigerant in the adsorption process. Thereby, the favourable characteristics of water as refrigerant can be retained while evaporation temperatures below 0 °C for freezing applications are rendered possible.

Experiments also show that increasing glycol mass fractions lead to lower heat flows in the evaporator. Lower heat flows are most likely due to two effects: decreased heat/mass transfer properties in the evaporator and a lower driving force. The driving force decreases due to lower partial water vapour pressures of the mixture in the evaporator. The heat/mass transfer properties are decreased due to the presence of glycol in the evaporator.

Using glycol as anti-freezing agent can, in particular, be of advantage for applications which require evaporation temperatures below 0 °C only occasionally. Glycol enables adsorption refrigeration at temperatures around the freezing point of pure water while retaining the good characteristics of water as refrigerant at higher temperatures.

To further increase the performance, the heat exchanger design of the evaporator can be adapted to the refrigerant mixture. Even without changing the heat exchanger design, retrofitting of existing adsorption chillers is possible: adding glycol to the evaporator can expand the operational temperature range for water adsorption applications below 0 °C.

Although the conducted experiments show that expanding the operational temperature below 0 °C is possible, some challenges remain for successful implementation in cooling applications. These challenges are discussed in the following.

Theoretical and experimental investigations at evaporation temperatures above 0 °C indicate that the addition of glycol has an almost negligible effect on the maximum achievable COP. Nevertheless, the maximum COP intrinsically decreases towards lower evaporation temperatures due to lower pressures in the evaporator during adsorption (as exemplarily shown in Figure 6.3). In addition, besides COP, there is another main performance indicator for adsorption chillers presented in the trade-off in the Pareto frontier in Figure 2.4: the specific cooling power SCP. The experiments indicate that the kinetics (heat/mass transfer properties) of the adsorption process relevant for the SCP also decrease with decreasing evaporation temperature and increasing glycol mass fraction. Consequently, the SCP will also decrease. As the SCP is important for economic operation of adsorption chillers, measures should be taken to increase the SCP. Restricting the glycol mass fraction to the minimal required mass fraction will most likely have a positive effect on heat and mass transfer in the additive-refrigerant mixture. Furthermore, heat transfer could be enhanced by improving the UA-value of the heat exchanger in the evaporator.

The presented investigation did not include a long-term evaluation of the anti-freezing agent glycol. Thus, important questions regarding long-term effects, e. g. “Does the anti-freezing agent slowly accumulate in the adsorbent or cause corrosion?” remain to be answered. In this work, no experimental evidence was found that would point to accumulation or corrosion, yet only long-term evaluation may provide definite answers required for successful implementation in a commercial product.

Liquid reflux from the condenser to the evaporator was challenging due to missing anti-freezing agent in the reflux. During the experiments shown in Figure 6.5, the liquid temperature in the evaporator fell below −3.5 °C during adsorption (cf. Figure E.1). Although the temperature of the liquid reflux from the condenser was approximately 35 °C, freezing did eventually occur at the inlet to the evaporator if the reduced pressure due to adsorption was present. Freezing was prevented by delaying the reflux from the condenser to the evaporator until the start of desorption: during desorption, the liquid temperature was around 0 °C and no freezing occurred at the evaporator inlet. However, such an operation is only possible in the investigated 1-bed configuration. Yet, for continuous cooling performance, usually, 2-bed configurations are employed, which continuously keep the liquid temperature in the evaporator below 0 °C. In the following, two possible solutions to prevent freezing in the reflux for 2-bed configurations are discussed.

The anti-freezing agent glycol could still be employed, if it is also present in the reflux. Since glycol does not participate in the adsorption process, it needs to be transferred to the reflux by other means. Due to the pressure difference, a pump is required to transport the non-freezing glycol–water mixture from the evaporator to the condenser. Such a system was also experimentally evaluated: in general, the evaluation was successful, yet circulation of the glycol–water mixture from the evaporator to the condenser also included transferring additional enthalpy from the condenser to the evaporator. Thus, less cooling power was provided by the system. Although improved insulation of the reflux to the surroundings could limit the enthalpy introduced into the evaporator, the power input and the general drawback of having a mechanical pump in the vacuum system will persist.

Alternatively, instead of a employing a pump, other types of anti-freezing agents could be employed to prevent freezing in the reflux: glycol stays in the evaporator and does not participate in the adsorption process, yet other additives (e. g. other alcohols) could be selected that participate in the adsorption process and, therefore, are also present in the reflux to prevent freezing. Selection of suitable additives, however, can become quite challenging as many requirements need to be considered: appropriate additives have to decrease the freezing point of the mixture and additionally not only evaporate but also participate in the adsorption process. Both process steps are known to be used for separation purposes and affect the concentrations of the additive in the liquid and vapour phase. Therefore, the vapour–liquid equilibrium as well as co-adsorption of the water–additive mixtures need to be assessed to determine the overall concentration of the additive required to fulfil the anti-freezing purpose in both the evaporator and the reflux.

Screening of water–additive vapour–liquid equilibria for suitable additives seems to be promising, although the scarcity of experimental isotherm data for co-adsorption will most likely require experiments or sophisticated predictive methods for assessment of most adsorbents. In general, water–additive mixtures with a suitable azeotrope (if found) could be chosen to avoid changes in concentrations of the additive due to evaporation. Non-azeotropic water–additive mixtures will either increase or decrease concentration during evaporation, and will, therefore, require higher overall concentrations of the additive compared to the concentration necessary to prevent freezing in the liquid of the evaporator or the reflux only. Alternatively, two additives could be combined: the first additive could be chosen to stay in the evaporator, e. g. glycol, and the second additive is chosen such that it strongly evaporates as well as ad- and desorbs, e. g. ethanol, to prevent freezing in the reflux. Finally, the overall concentration of the additive should be as low as possible in order to minimise the deterioration of the outstanding properties of the refrigerant water for the adsorption process, e. g. high enthalpy of vaporisation. In the end, for water-based refrigerant mixtures to prevail against other pure working pairs in the same temperature range, as many advantages of the natural refrigerant water as possible will need to be retained.

Chapter 7

Summary, conclusions and future perspectives

This thesis focuses on the evaporator of water-based adsorption chillers and addresses two bottlenecks of the natural refrigerant water:

(I) The challenge of efficient, sub-atmospheric evaporation of water to achieve high heat transfer coefficients

(II) The limitation towards lower temperatures in the evaporator due to freezing of water

In this chapter, the contributions of this thesis are summarised and the main conclusions are given in Section 7.1. Possible future works are discussed in Section 7.2. Both sections are clustered according to the bottlenecks (I) and (II).

7.1 Summary and conclusions

(I) Capillary-assisted, sub-atmospheric evaporation of water

The low saturation pressure of the natural refrigerant water hinders efficient evaporation at low temperatures: high heat transfer coefficients by nucleate boiling can only be achieved with high superheats. High superheats, however, reduce the thermodynamic efficiency of adsorption chillers. Efficient evaporation of water at low temperatures is enabled by thin-film evaporation. Thin refrigerant films can be created on heat exchanger surfaces without additional pumps by capillary action. Capillary-assisted, sub-atmospheric, thin-film evaporation of water is studied in this thesis and the following main conclusions are derived

Impacts of experimental setups on measuring sub-atmospheric, thin-film evaporation of water

In cooperation with the Fraunhofer ISE, important factors were identified to obtain comparable results of sub-atmospheric, thin-film evaporation experiments. Identified factors included (I) the assessment and consideration of uncertainty in measurement, (II) the importance of identical experimental input conditions, (III) the exact definition of the investigated heat exchanger's surface properties and (IV) control of non-condensable gases. All factors are considered in the experimental setup employed in this thesis and should also be paid attention to in future investigations.

Validation of the experimental setup

The experimental setup and procedure for evaporation experiments were successfully validated: the reproducibility of results from repeated automated experiments was shown to be very good. Furthermore, the comparison with identical measurements conducted at an experimental setup at Fraunhofer ISE showed generally good agreement for higher driving forces and identical results within uncertainty of measurement for low driving forces. The impact of the driving force for the comparison was not fully understood and calls for further investigations. Nevertheless, the employed experimental setup and procedure have proven to generate reproducible results and are well suited to conduct investigations regarding sub-atmospheric, thin-film evaporation of water.

Experimental proof: decreasing filling levels are well-suited to characterise capillary-assisted, thin-film evaporation at all filling levels

Experiments with decreasing filling levels were compared to experiments at constant filling levels for capillary-assisted, thin-film evaporation of water. The results showed the same characteristics and measured differences were close to or within the standard uncertainty of measurement. Thus, experiments with continuously decreasing filling level are shown to be well-suited for quickly characterising heat exchangers: within a single experiment, the overall heat transfer coefficient can be determined for all filling levels. Although the determined differences were close to or within the standard uncertainty of measurement, the overall heat transfer coefficients were generally 5-10 % higher in experiments with continuously decreasing filling levels than in experiments at constant filling level. The results point to an underlying physical effect. Thus, operating thin-film evaporators with decreasing filling level and only intermittent refills could increase the maximum achievable U-value by up to 10 %.

Coatings increase the UA-value by up to a factor of 10

Maintaining a thin, evaporating refrigerant film on the heat exchanger's surface can be realised by capillary action. Capillary action is governed by interactions of the refrigerant and the properties of the heat exchanger surface, which can be tailored by applying coatings. The evaporation performances of 7 coatings and a plain tube were experimentally determined for all filling levels at different temperatures and driving forces to improve the understanding of capillary-assisted, sub-atmospheric, thin-film evaporation of water. The experimental results show that the investigated coatings can increase the UA-value compared to plain tubes by a factor of up to 10. In general, increasing evaporator inlet temperatures and lower filling levels increase UA-values. UA-values increase for lower filling levels because the area exploited by thin-film evaporation increases, yet only as long as no dry-out occurs.

Dry-out of tubes limits the evaporation performance and explains inconclusive findings regarding the driving force in literature

Dry-outs severely limit the evaporation performance: similar to the dry-out described in heat pipes, dry-out of the coated tubes' surface occurs at high evaporation rates. As evaporation rates increase with higher driving forces and lower filling levels, these operating conditions are prone to lead to dry-out and should, therefore, be used to evaluate the performance of thin-film evaporation. In the conducted experiments, the best performing coating did not suffer from dry-out.

As long as no dry-out occurred, no impact of the driving force on the UA-value could be detected. Divergent findings in literature regarding the impact of the driving force could be explained if dry-out also occurred in these experiments. Dry-out can be prevented, if the capillary action is strong enough: continuous wetting by capillary action is governed by the interaction of refrigerant and coating. Therefore, to keep the tube wetted even at high driving forces, the coating's characteristics are pivotal.

Characteristic properties of coatings to keep tubes wet: a guideline for future evaporator development

The evaporation performance was correlated to 4 surface properties of the investigated 7 coatings. Thereby, a guideline for future evaporator development could be derived: coatings should be tailored to increase surface roughness, extension and porosity. Thicker layers and closed porosity are probably not beneficial for evaporation performance. Thus, to minimise costs and increase thermal conductivity, coatings should be as thin as possible and closed pores avoided.

(II) Water-based adsorption cooling below 0 °C

Cooling with water-based adsorption chillers is restricted to temperatures above 0 °C due to freezing of the refrigerant. The freezing point of the refrigerant water can be lowered by adding the anti-freezing agent glycol to the evaporator of the adsorption chiller. The experimental assessment of water-based adsorption cooling below 0 °C in this thesis resulted in the following main conclusions.

Anti-freezing agent glycol overcomes limitation of the natural refrigerant water: water-based adsorption cooling below 0 °C

Addition of the anti-freezing agent glycol to the evaporator renders water-based adsorption cooling below 0 °C possible. Glycol prevents freezing in the evaporator and the experimental results indicate that only water with its favourable properties serves as refrigerant in the adsorption process. Hence, by simple addition of glycol to the evaporator, even existing adsorption chillers could be retrofitted to expand their operational range below 0 °C.

Addition of glycol particularly promising for occasional operation around 0 °C

Both the theoretical and experimental analysis show that the addition of glycol only slightly affects the efficiency (COP) of the process. However, the COP intrinsically decreases towards lower evaporation temperatures. Furthermore, the specific cooling power (SCP) strongly decreases due to the addition of glycol. Most likely, the reasons are decreased heat / mass transfer properties in the evaporator as well as a lower driving force. As a result, mass fractions of glycol should be chosen as low as possible to still prevent freezing. Additionally, the heat transfer in the evaporator should be improved for successful implementation of water-based adsorption cooling below 0 °C. Employing the anti-freezing agent glycol can be particularly promising for applications which only occasionally require temperatures around 0 °C: while a low mass fraction of glycol prevents freezing at low temperatures, the performance and favourable characteristics of the refrigerant water are mostly retained for operation at higher evaporation temperatures.

7.2 Future perspectives

Based on the experience gained during the work for this thesis, the following topics have been identified as possible subjects for future research.

(I) Capillary-assisted, sub-atmospheric evaporation of water

Balancing inner and outer heat transfer – exergetic analysis of the heat transfer process

Sub-atmospheric, thin-film evaporation greatly improves the heat transfer for the phase change on the heat exchanger's outer surface. This surface is usually heated by a secondary circuit from the inner side of the heat exchanger involving another possibly limiting heat transfer. To balance the overall heat transfer through the heat exchanger, improvement of the outer heat transfer due to thin-film evaporation needs to be supported by improvement of the inner heat transfer as well. In this context, enhancing the inner heat transfer often implies introducing turbulence structures in the liquid phase which tend to cause pressure drops. Thus, to reflect required pumping power, enhancements should follow a holistic design approach and also incorporate an exergetic analysis of the heat transfer process.

Enhancing the experimental setup for improved comparability and further analysis

To compare results from evaporation experiments, identical experimental conditions are crucial. The latter can be ensured by controlling the inlet temperatures and volume flows as well as evaporator pressure and filling level. In the employed experimental setup, only the inlet temperatures were actively controlled by thermostats. In contrast, volume flows as well as evaporator pressure and filling level were manually set in an iterative process. Hence, automatic control would drastically reduce the required experimental time and improve the reproducibility of conducted experiments. Necessary control loops could be realised by implementing online detection of the filling level as well as advanced communication to the thermostats to control the pumps. The pressure in the evaporator could be controlled by the condenser inlet temperature or volume flow to enable experiments at constant pressure or constant driving force. Furthermore, the impact of ambient conditions on the experiments should be minimised by encapsulating the setup and controlling both surrounding temperature and humidity.

Increased quality of the photos taken of the investigated tubes during the experiment could be realised with a macro lens and could open up new possibilities for evaluation. For macroscopic structures (e. g. finned tubes), contact angles between the fins could possibly be resolved and, thus, analysed in the course of the experiments. For microscopic structures (e. g. coatings), wetting of the surface could be analysed in detail and dry-out possibly detected automatically by image detection.

Pushing the boundaries of capillary-assisted evaporation

Following the derived guideline for coating development in this thesis, coatings should be tailored to increase surface roughness, extension and porosity. As the best performance was measured at the highest values of these surface characteristics, additional improvements can be expected for further increased surface roughness, extension and porosity.

In addition to the assessment of the surface characteristics evaluated in this thesis, methods should be derived to determine and compare contact angles at process conditions on macroscopically non-flat surfaces, too. Analysing the refrigerant film thickness during the experiment would also provide valuable information for the understanding of thin-film evaporation from coated tubes as the thickness is important in the context of evaporation from extended menisci.

Coatings could also be applied to heat exchanger designs beyond tubes, e. g. compact plate heat exchangers. Vacuum-proof designs of plate heat exchangers could potentially provide advantages regarding cost, scalability and utilisation of available space. Finally, the effect of transient operation, e. g. decreasing filling levels, shows promising results, calling for further investigation to possibly even improve existing evaporator designs by simply adjusting control.

Understanding evaporation mechanisms in capillary-assisted evaporation

Future investigations could be targeted at understanding the underlying evaporation phenomena in the vicinity of the 3-phase contact line in coatings or precisely defined macroscopic evaporation structures. Knowledge of shape, size and length of menisci could also help to further improve the evaporation performance and precisely tailor structures for sub-atmospheric evaporation – possibly even tailored for refrigerant mixtures. In addition, better understanding of the involved evaporation mechanisms could also be beneficial for other applications of thin-film evaporation, e. g. heat pipes and nucleate boiling.

(II) Water-based adsorption cooling below 0 °C

The results of the conducted experiments indicate that glycol remains in the evaporator and does not participate in the adsorption process. Hence, no anti-freezing agent is present in the liquid reflux from the condenser to the evaporator. To prevent freezing of the reflux, additional measures need to be taken. For the investigated 1-bed adsorption chiller, reflux during adsorption was delayed to the desorption phase in which higher temperature and pressure in the evaporator prevented freezing. For continuous operation of the evaporator in common 2-bed configurations, however, other solutions are required to prevent freezing. In the following, 2 possible solutions are presented to overcome the challenge of freezing in the reflux.

Challenge to prevent freezing in the reflux – extending the anti-freezing effect of glycol into the reflux

Freezing in the reflux can be prevented by circulating the non-freezing glycol–water mixture from the evaporator to the condenser with a pump. Operating a pump in a vacuum system to circulate a saturated liquid is challenging because of cavitation; yet, the circulation of the glycol–water mixture would also open up possibilities for enhanced heat transfer by falling-film evaporation. The design of the circulation system would need to ensure that introduction of unwanted enthalpy from the condenser into the evaporator is minimised to not reduce cooling power. Experimental evaluation of such a system could contribute to successful implementation of water-based adsorption cooling below 0 °C.

Challenge to prevent freezing in the reflux – screening suitable additives beyond glycol

Freezing in the reflux could also be prevented by selecting other additives which evaporate and participate in the adsorption process. For the selection of suitable additives, several requirements need to be considered: besides lowering the freezing point of the mixture with water, the additives need to evaporate, ad- and desorb and should only be required in low overall mass fractions to retain the positive characteristics of water as the refrigerant.

Additives could be evaluated regarding these requirements in a screening process: solid–liquid (freezing point depression) and vapour–liquid (vapour phase for co-adsorption) data could be assessed to identify potential additives. Next, as experimental data of co-adsorption is scarce, co-adsorption for promising additives would need to be determined by experiments or predictive methods. Thereby, promising additives beyond glycol could be identified systematically and subsequently be experimentally evaluated to enable water-based adsorption cooling below 0 °C.

(I) and (II): Towards tailored, capillary-assisted, sub-atmospheric, thin-film evaporation for refrigerant mixtures

A major challenge for the successful implementation of water-based adsorption cooling below 0 °C with the anti-freezing agent glycol is the improvement of heat transfer. Heat transfer for sub-atmospheric evaporation of pure water is challenging. Moreover, heat transfer coefficients can additionally be considerably smaller for evaporation of mixtures (e. g. water–glycol) compared to evaporation of pure substances (Baehr and Stephan, 2011).

In this thesis, capillary-assisted, thin-film evaporation is studied to increase heat transfer coefficients for sub-atmospheric evaporation of pure water. Evaporation of water–additive mixtures has not been the focus of the conducted evaporation experiments. However, the evaporator of the adsorption chiller was capillary-assisted to achieve high heat transfer coefficients: the axially-finned MicroFin tubes created capillary action between their fins with pure water. Unfortunately, heat transfer with this finned tube was no longer reliable upon addition of glycol: the addition of glycol probably affected the capillary action resulting in a rapid decrease of the effective heat transfer area. Consequently, the heat flow in the evaporator reproducibly broke down. To maintain high heat transfer coefficients upon the addition of the anti-freezing agent, capillary-assisted evaporators probably have to be tailored to the anti-freezing agent–water mixture employed.

In general, capillary-assisted evaporation of additive–water mixtures will most likely face challenges connected to the phenomenon of concentration polarisation. Concentration polarisation occurs as a consequence of mixture evaporation. If an additive tends to evaporate less than water, the concentration of the additive locally increases at the place of evaporation, thereby creating additional diffusion resistance. However, the place of evaporation is, at the same time, the place of minimal temperature in the evaporator. Therefore, a locally increased concentration of the additive can be beneficial to reliably prevent freezing. Nevertheless, diffusion resistance could possibly dominate the overall process and hinder evaporation of water from the tube.

Mechanical measures could help to reduce concentration polarisation on the tube's surface: the refill process of refrigerant from the condenser to the evaporator can create a splash. The evaporator could be designed in such a way that the splash is exploited to wash surplus additive from the thin-film evaporation surface and create a more homogenous distribution of the additive in the evaporator.

Although challenges for thin-film evaporation of refrigerant mixtures remain, the contributions of this thesis provide a good basis to tap the potential of capillary-assisted evaporators for refrigerant mixtures and enable water-based adsorption cooling below 0 °C.

Appendices

Appendix A

Overview of evaporator patents

In this appendix, a list with 18 patents containing heat exchangers suitable for evaporators of adsorption chillers is given in Table A.1.

Table A.1: Published patents involving evaporator heat exchanger in order of publication date.

Short description	Evaporator heat exchanger type	Patent applicant or assignee	Reference
Process for manufacturing an evaporator/condenser	Double-walled metal sheet	Vaillant GmbH	(Lang et al., 2001) DE10053045A1
Evaporator for adsorption heat pump or refrigerator	Lamella heat exchanger	Bosch Thermo-technik GmbH	(Faust, 2001) DE10011688A1
Absorption chiller including an evaporator	Falling-film shell-and-tube heat exchanger	Ph. SonnenWärme AG & ZAE Bayern	(Costa et al., 2006) DE102004039327A1
Capillary evaporator for use in e. g. heat pump	Capillary active structured tube	SorTech AG	(Mittelbach and Wondra, 2009) DE102008028854A1
Device/prod. method for absorbing by capillarity	Capillary structure on folded metal sheet	Mahle Behr Gm-bH und Co. KG	(Burk and Zwittig, 2010) EP1918668B1
Evaporator for sorption machines	Tubes surrounded by capillary active fibres	InvenSor GmbH	(Braunschweig and Paulussen, 2011) DE102009053843A1
Evaporator with multiple supply pipes	Plate with capillary active structure	TU Darmstadt	(Schweizer et al., 2011) DE102010016644A1
Module for a heat pump with evaporator/condenser	Flat tube with various cross sections	Behr GmbH and Co. KG	(Burk et al., 2013) DE102011079536A1
Method for alternating evaporation & condensation	Modified heat transfer area, e. g. tubular/flat	Fahrenheit GmbH	(Mittelbach and Dassler, 2014) EP2689201B1

...continued on next page

Table A.1 – continued from previous page

Short description	Evaporator heat exchanger type	Patent applicant or assignee	Reference
Evaporator/prod. method for exhaust line in car	Tube in tube with sintered wire mesh	Benteler Automobiltechnik GmbH	(Rubitschek et al., 2014) DE102013103840A1
Evaporator heat exchanger	Pipe coil surrounded by metal wire mesh	Vaillant GmbH	(Spahn and Szuder, 2016) DE102014223250A1
Evaporator for adsorption heat pump	Powder coated pipe coil	Vaillant GmbH	(Spahn et al., 2016) DE102014224137A1
Surface feeding & distribution of refrigerant	Enhanced tube/channel with porous materials	InvenSor GmbH	(Braunschweig et al., 2016) EP2473811B1
Heat exchanger for an evaporator	Enhanced tube with sintered metal coating	Vaillant GmbH	(Spahn and Szuder, 2017) DE102015213320A1
Evaporator for refrigerant circuits in cars	Tube on porous sponge (to reduce vibrations)	BMW AG	(Friedrich, 2017) DE102015220895A1
Adsorption air-conditioning system	Plate-and-tube with phase change materials	Ford Global Techn. LLC	(Levin et al., 2017) US9789746B2
Evaporator/capacitor with porous particles	Enhanced porous surface, e. g. with tubes	Fraunhofer Gesellschaft	(Baumeister and Weise, 2017) DE102016209082A1
Using splashing by bubbles to wet hx	e. g. enhanced (finned) tube/plate-and-tube	Fraunhofer Gesellschaft	(Warlo et al., 2018) WO2018033418A1

Appendix B

Details of the experimental setups

In this appendix, additional information about the experimental setups used in this thesis is provided. Both setups, including the components, hydraulic circuits, data acquisition and control, were developed, custom-built and assembled at the Chair of Technical Thermodynamics, RWTH Aachen University.

Details regarding the experimental setup for the adsorption experiments are given in Section B.1, regarding the evaporation experiments in Section B.2. Section B.3 summarises experimental equipment used in both setups.

B.1 Adsorption chiller

In this section, additional information about the experimental setup employed for experiments with the anti-freezing agent ethylene glycol in Chapter 6 are given. The adsorption chiller consisted of 3 main components: adsorber, evaporator and condenser (cf. Figure B.1). Details for each component are given in the next Sections B.1.1 to B.1.3. Afterwards, details regarding the experimental equipment is provided in Section B.1.4.

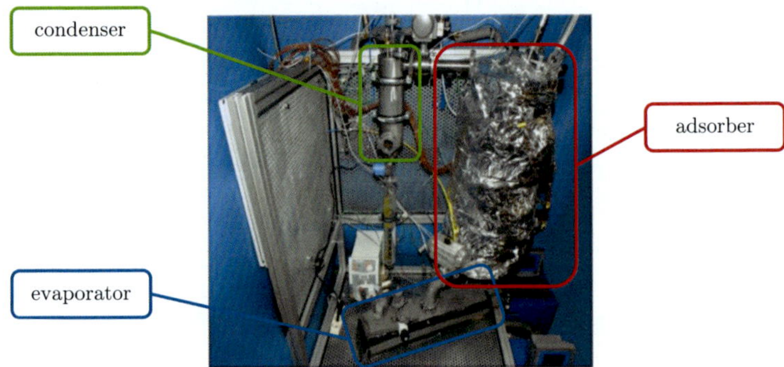

Figure B.1: Photo of the experimental setup of the adsorption chiller.

B.1.1 Adsorber

Geometric details of the adsorber vessel, insulation and installed heat exchanger (see Figure B.2) are presented in Table B.1. A detailed description of the adsorber is given in Section 3.2.2.

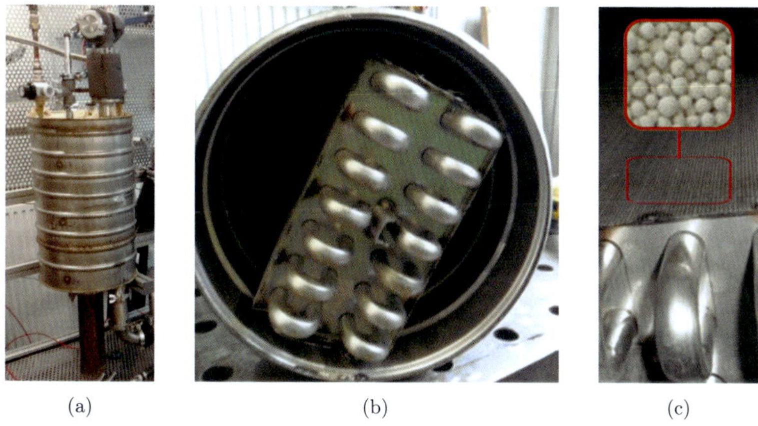

(a) (b) (c)

Figure B.2: Adsorption chiller: photos of the adsorber. Total view of cylindric adsorber vessel (a), bottom view on the heat exchanger in open vessel (b) and close-up of fin-coil heat exchanger covered with vapour-permeable, stainless-steel sheeting and granular zeolite NaY (c).

Table B.1: Adsorption chiller: adsorber specifications for vessel, insulation, heat exchanger and adsorbent. Photos of the adsorber are shown in Figure B.2.

Parameter	**Value**
Vessel: cylindrical	
Outer diameter	350 mm
Height	550 mm
Insulation	
Material	Glass wool with aluminium foil
Thickness	120 mm
Heat exchanger: fin coil	
Material	Stainless steel
Number of serial passages	28
Adsorbent: granular zeolite NaY	
Dry adsorbent mass	About 5 kg filled between fins (lamella)
Cover to keep adsorbent in place	Vapour-permeable, stainless-steel sheeting

B.1.2 Evaporator

Geometric details of the evaporator vessel, insulation and installed heat exchanger (see Figure B.3) are presented in Table B.2. A detailed description of the evaporator is given in Section 3.2.2.

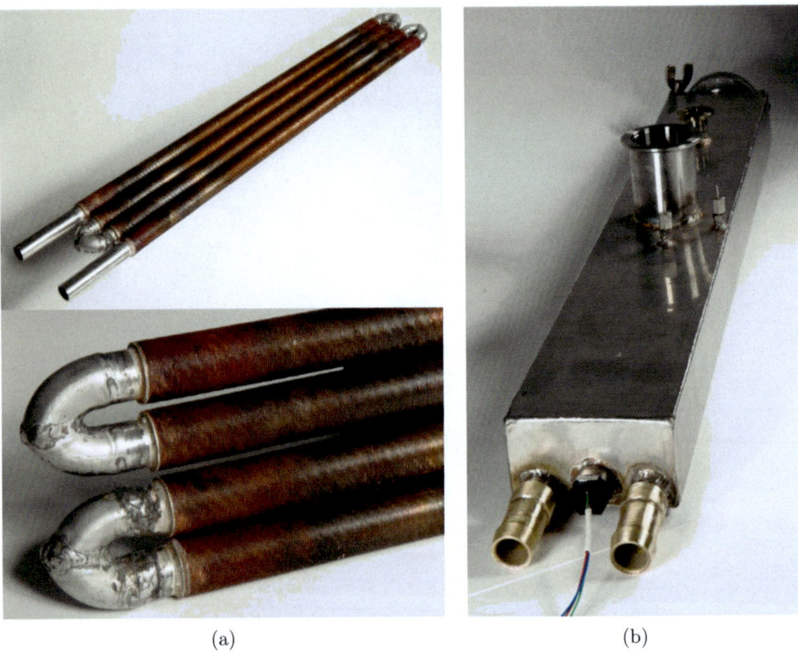

(a) (b)

Figure B.3: Adsorption chiller: photos of the evaporator. Close-up and total view of 4 serially connected "MicroFin" tubes with in- and outlet as well as "U-bend" (a). Evaporator vessel with in- and outlet of secondary circuit and binary filling level sensor in between (b). Some ISO-KF flanges actually present on the evaporator cannot be seen on this photo.

Table B.2: Adsorption chiller: evaporator specifications for vessel, insulation and finned-tube heat exchanger. Photos of the evaporator are shown in Figure B.3.

Parameter	**Value**
Vessel: rectangular-shaped cuboid	
Length	600 mm
Width	85 mm
Height	45 mm
Insulation	
Material	Conel Flex EL Isolierplatte CLIEL25
Thickness	25 mm
Heat exchanger: finned tube	
Manufacturer	Schmöle GMBH
Type	Trufin S/TT 09-40 1710013
Material	Cu
Number of tubes	4
Length of each tube	500 mm
Overall tube length (with U-bends)	2310 mm
Overall finned length	2000 mm
Tube inner diameter	15 mm
Tube outer diameter (fin tips)	18.8 mm
Fin spacing	40 fpi
Fin width	0.3 mm
Fin height	0.9 mm
Inner turbulence structure	Yes

B.1.3 Condenser

Geometric details of the condenser vessel, insulation and installed heat exchanger (see Figure B.4) are presented in Table B.3. A detailed description of the evaporator is given in Section 3.2.2.

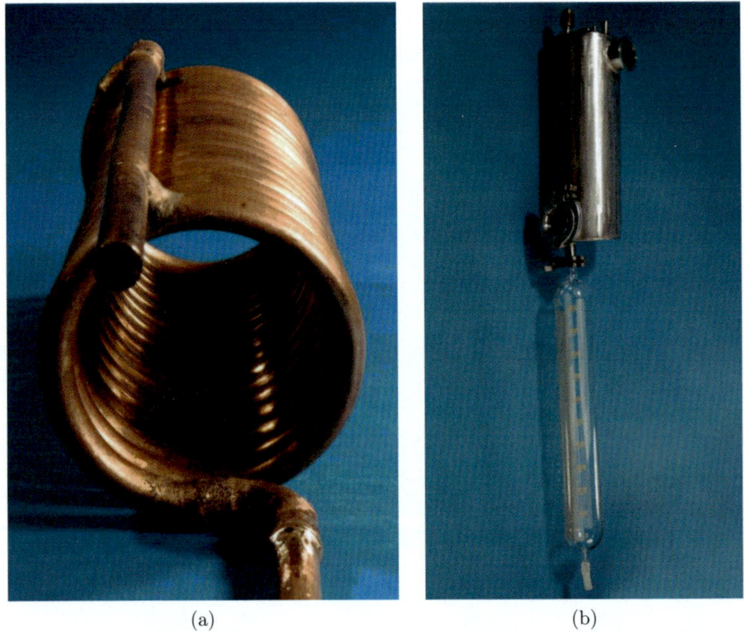

(a) (b)

Figure B.4: Adsorption chiller: photos of the condenser. Coiled copper tube (a), condenser vessel with viewing window and ISO-KF flanges and attached glass measurement cylinder (b).

Table B.3: Adsorption chiller: condenser specifications for vessel, insulation and heat exchanger coiled tube. Photos of the condenser are shown in Figure B.4.

Parameter	Value
Vessel: cylindrical	
Outer diameter	110 mm
Height	355 mm
Insulation	
Material	Glass wool with aluminium foil
Heat exchanger: coiled tube	
Manufacturer	Self-built
Material	Cu
Type	Helix
Number of turns in helix	17.5
Average diameter of helix	75 mm
Gradient of helix per turn	14 mm
Length of coiled tube	4130 mm
Overall tube length (incl. supply line etc.)	4550 mm
Tube inner diameter	8 mm
Tube outer diameter	10 mm

B.1.4 Experimental equipment

The experimental setup employed to characterise adsorption chillers is described in detail in Section 3.1. In the following, additional information regarding the experimental equipment is provided.

Table B.4 provides a summary of all valves and pumps installed in the experimental setup. The first 3 listed types of valves are used in the adsorption chiller and are, therefore, designed for vacuum conditions. All electric vacuum valves are included in Figures 3.5 and 6.4; the manual vacuum valves are not included in the figures. The remaining hydraulic valves are installed in the secondary circuits of the components and are also included in the according layouts in Figures 3.4 and 3.14. The peripheral pumps are part of the secondary circuit of the adsorber given in Figure 3.4.

In Table B.5, sensors necessary to reliably operate the adsorption chiller are summarised. The pressure transducers were used to determine the pressure in each component. The pressure transducers in the evaporator and condenser allowed to detected the presence of non-condensable gases by comparing their readings to the saturation pres-

Table B.4: Experimental equipment: used valves and pumps with manufacturer, type and location in experimental setup. Relevant switching status according to the phases of the adsorption chiller cycle are given in Table 3.2.

Type	**Manufacturer**	**Model**	**Location**
Valves			
direct-acting plunger valve	Riegler & Co. KG	220V vacuum-tight	Reflux below condenser
Pneumatic butterfly valve	Schwarz Vakuum-technik GmbH	E03, DN40 with 230V AC pilot valve	Steam adsorber to evaporator/condenser
Manual angle valve	Leybold GmbH or Pfeiffer Vacuum	ISO-KF DN16 Al/VA	All components to surroundings
Inclined-seat valve	Bürkert Fluid Control Systems	Typ 2000, A13	8x in hydraulic circuit
Manual hand valve	Bürkert or Conex Bänninger	standard ball valve	5x in hydraulic circuit
Safety relief valve	Goetze KG Armaturen	851 b GF 1/2" 15 bar	2x in hydraulic circuit
Manual control valve	Bürkle GmbH	3/2 directional, PVDF, 13 mm, DN8	Before evaporator/ condenser in- and outlet
Manual control valve	Bürkert or Conex Bänninger	3/2 directional, metal	Before adsorber in- and outlet
Pumps			
Peristaltic pump	Heidolph GmbH & CO. KG	Pumpdrive 5101	Between condenser and evaporator
Peripheral pump	Speck Pumpen Systemtechnik GmbH	NPY-2251-MK-TOE-HT	Adsorber hot hydraulic circuit
Peripheral pump	Speck Pumpen Systemtechnik GmbH	NPY-2051	Adsorber cold hydraulic circuit

sure of the measured temperature in the respective liquid phases of the components. If non-condensable gases were present, they usually accumulated in the condenser during operation of the adsorption chiller and were removed with a vacuum pump (Table B.11). The binary filling level sensor was installed in the evaporator and used to maintain a constant filling level when automated refilling was activated.

Table B.5: Experimental equipment: employed pressure transducers and filling level sensor for control of adsorption chiller with manufacturer, type and location in experimental setup.

Type	**Manufacturer**	**Model**	**Location**
Pressure transducer (3x)	Keller AG, CH	PAA-21S/80549.3	Adsorber, evaporator and condenser
Filling level sensor (binary)	First Sensor AG	OLP Series	Evaporator between in- and outlet

Operating the experimental setup was realised with the software LabVIEW. The implemented LabVIEW program was a timed state machine and allowed to set the temperatures of the thermostats, control the states of the valves, read data from the sensors, apply calibrations to the sensor readings, calculate reduced quantities and store all data in a text file. The text files included all sensors readings with and without applied calibrations, set temperatures and status of the valves and pumps, time stamps and the according phase of the adsorption chiller. The data was collected at a frequency of 1 Hz. The front panel of the LabVIEW program (Figure B.5) displayed plots of all sensor data and valve positions necessary for operating the experimental setup and monitoring the experiments.

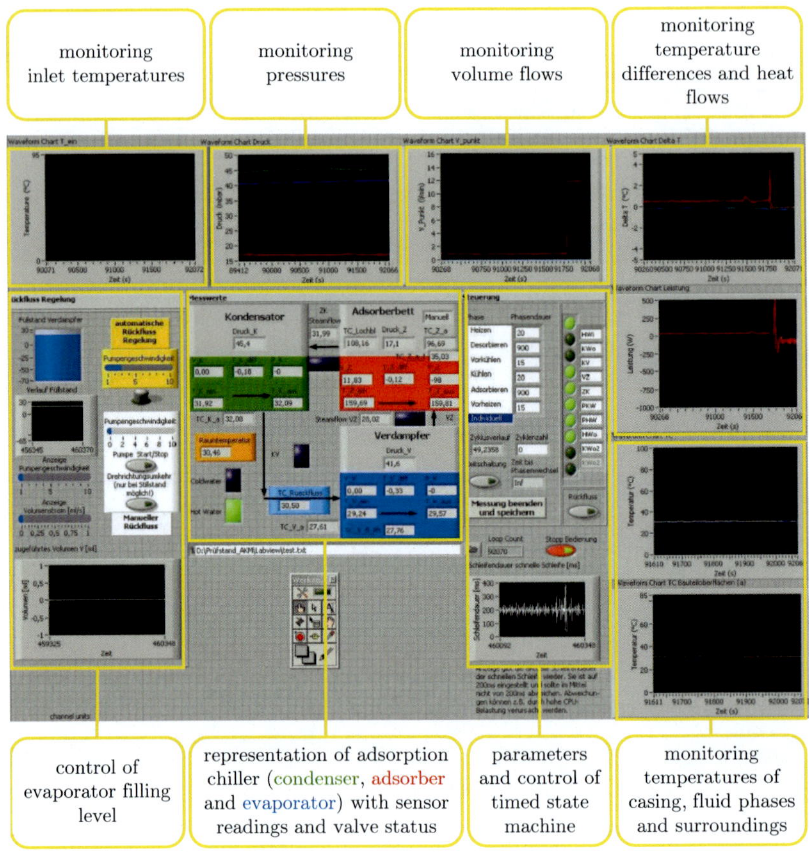

Figure B.5: Adsorption chiller: screenshot of the front panel of the LabVIEW program implemented to operate the adsorption chiller experimental setup.

B.2 Evaporation experiments

In this section, additional information about the experimental setup employed for the sub-atmospheric, thin-film evaporation experiments in Chapters 4 and 5 are given. The experimental setup consisted of 2 main components: evaporator and condenser (cf. Figure B.6). Details for both components are given in the next Sections B.2.1 and B.2.2. Afterwards, details regarding the experimental equipment is provided in Section B.2.3.

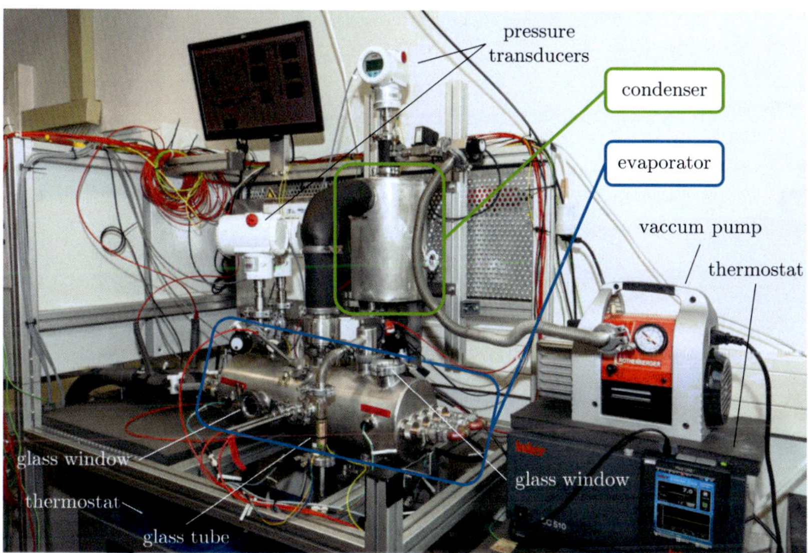

Figure B.6: Photo of the setup for the evaporation experiments during assembly. The evaporator and condenser as well as the external parts of the tubes (extending on both sides of the evaporator) are not yet insulated.

B.2.1 Evaporator

Geometric details of the evaporator's vessel (Figure B.7), insulation and suitable tubes to be investigated in the experimental setup for the evaporation experiments are presented in Table B.6. A detailed description of the evaporator is given in Section 3.2.2.

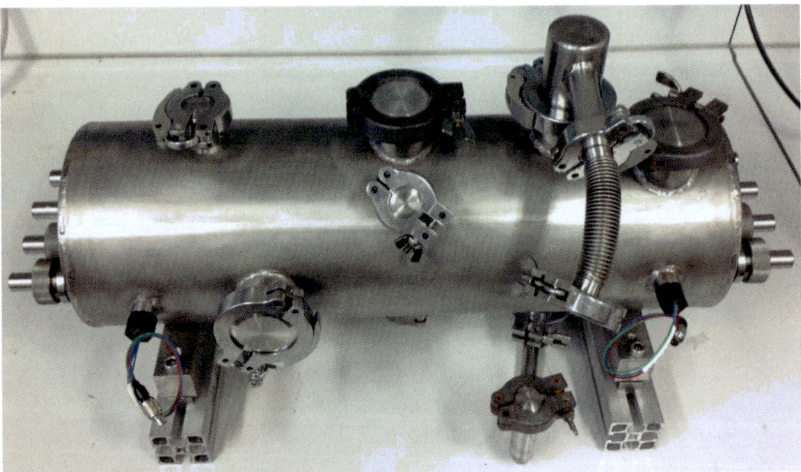

Figure B.7: Photo of the evaporator vessel for the evaporation experiments before installation with ISO-KF flanges protected by blind flanges, 2 installed binary filling level sensors and short, flexible, corrugated hose for installation of the glass tube. Geometric details of the vessel, insulation and suitable tubes are given in Table B.6.

Table B.6: Experimental setup evaporation: evaporator specifications for vessel, insulation and suitable tubes.

Parameter	Value
Vessel: cylindrical	
Length	511 mm
Outer diameter	168.3 mm
Wall thickness cylindrical surface	3 mm
Wall thickness front and back	4 mm
Insulation	
Material	Conel Flex EL Isolierplatte CLIEL25
Thickness	25 mm
Heat exchanger: suitable tubes	
Number of tubes	4
Length of each tube in evaporator	500 mm
Total length of each tube	~650 mm
Max. outer tube diameter	15 mm

B.2.2 Condenser

Geometric details of the condenser's heat exchanger (Figure B.8 (a)), vessel (Figure B.8 (b)) and insulation used in the eperimental setup for the evaporation experiments are presented in Table B.7. A detailed description of the condenser is given in Section 3.2.1.

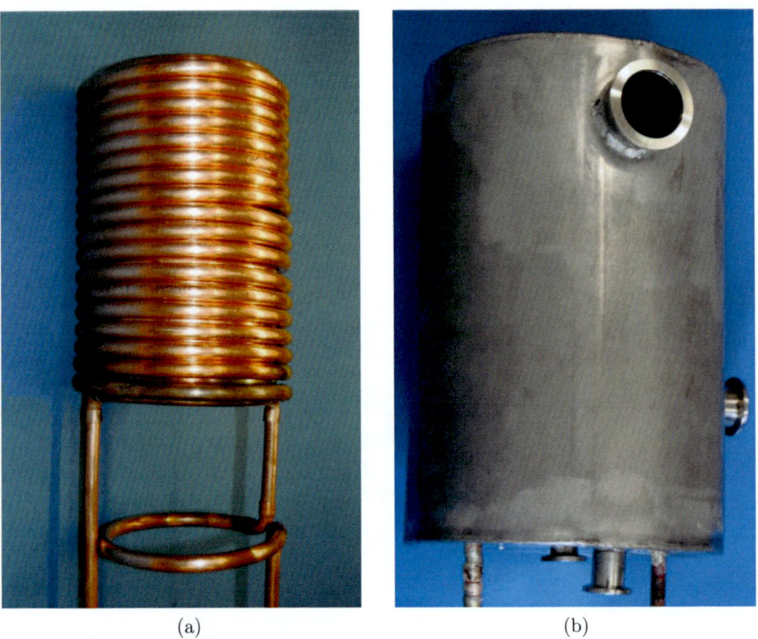

(a) (b)

Figure B.8: Photo of the coiled-copper-tube double-helix used as heat exchanger in the condenser (a) and the condenser vessel with ISO-KF flanges and hydraulic connections for the heat exchanger (b). The front ISO-KF flange on the bottom extends a little further below the ISO-KF flange behind to enable fastening of ISO-KF clamps on both shown ISO-KF flanges at the same time. Geometric details of the heat exchanger, vessel and insulation are given in Table B.7.

Table B.7: Experimental setup evaporation: condenser specifications for vessel, insulation and heat exchanger coiled tube.

Parameter	Value
vessel: cylindrical	
Outer diameter	200 mm
Height	306 mm
Insulation	
Material	Conel Flex EL Isolierplatte CLIEL25
Thickness	25 mm
Heat exchanger: coiled tube	
Manufacturer	Self-built
Material	Cu
Type	Double-helix
Number of turns in outer helix	18 + 1 below
Number of turns in inner helix	~ 20
Length of coiled tube	13 950 mm
Tube inner diameter	8 mm
Tube outer diameter	10 mm

B.2.3 Experimental equipment

The experimental setup employed to characterise sub-atmospheric, thin-film evaporation is described in detail in Section 3.2. In the following, additional information regarding the experimental equipment is provided.

In Table B.8, filling level sensors installed in the evaporator are summarised. However, only information obtained from the image recognition (cf. Section 3.2.2) was used to determine the filling levels in the conducted experiments. Other installed filling level sensors did not reliably detect the filling level at all operating conditions. Yet, these sensors were helpful for operating the setup since the filling level obtained by image recognition was only available after the experiments were finished and evaluated.

Table B.9 provides a summary of all valves and pumps installed in the experimental setup. The first 4 listed types of valves are designed for vacuum conditions. All electric vacuum valves except the valve to the vacuum pump are included in Figures 3.6 and 5.1; the manual vacuum valves are not included in the figures. The remaining hydraulic valves are installed in the secondary circuits of the components and are included in the layout in Figure 3.14.

Table B.8: Experimental equipment: installed filling level sensors for detection of filling level in evaporator with manufacturer, type and location in the evaporator of the experimental setup.

Type	**Manufacturer**	**Model**	**Location**
Filling level sensor (binary)	First Sensor AG	OLP Series optical detection	2x upper & 2x lower edge of tube
Filling level sensor (continuous)	First Sensor AG	CLC Series capacitive detection	Glass tube outside of evaporator
Image recognition digital camera	Canon	EOS 1300D (18 MP) EF-S 18-55 mm IS	Glass tube outside of evaporator

Table B.9: Experimental equipment: used valves with manufacturer, type and location in experimental setup.

Valve type	**Manufacturer**	**Model**	**Location**
Direct acting valve valve	Bürkert Fluid Control Systems	6027 A 6,0 FKM VA 24DC NA01	Reflux below condenser & reflux before evaporator
Direct acting plunger valve	Bürkert Fluid Control Systems	6013 A 6,0 FKM VA 24DC	Between condenser and vacuum pump
Pneumatic butterfly valve	Schwarz Vakuumtechnik GmbH	high-vacuum DN40 with 24V DC pilot valve	Steam flow between evaporator condenser
Manual angle valve	Schwarz Vakuumtechnik GmbH	high-vacuum DN16 VA	Evaporator to surroundings & condenser to surroundings
Manual control valve	Bürkle GmbH	3/2 directional, PVDF, 13 mm, DN8	Before components in- and outlet

Operating the experimental setup was realised with the software LabVIEW. The implemented, object-oriented LabVIEW program was a timed state machine and allowed to set the temperatures of the thermostats, control the states of the valves, trigger the camera and save taken pictures with a time stamp, read data from the sensors, apply absolute and relative calibrations to the sensor readings, calculate reduced quantities and store all data in a text file. The text files included all sensors readings with and without applied calibrations, set temperatures and status of the valves and pumps, time stamps and the according phase of the setup. The data was collected as often as the data could be read from the sensors in order to allow for accurate control of the setup. Synchronised, collected data was saved to the text file at a frequency of 1 Hz. The front panel of the LabVIEW program (Figure B.9) displayed plots of all sensor data and valve positions necessary for operating the experimental setup and monitoring the experiments.

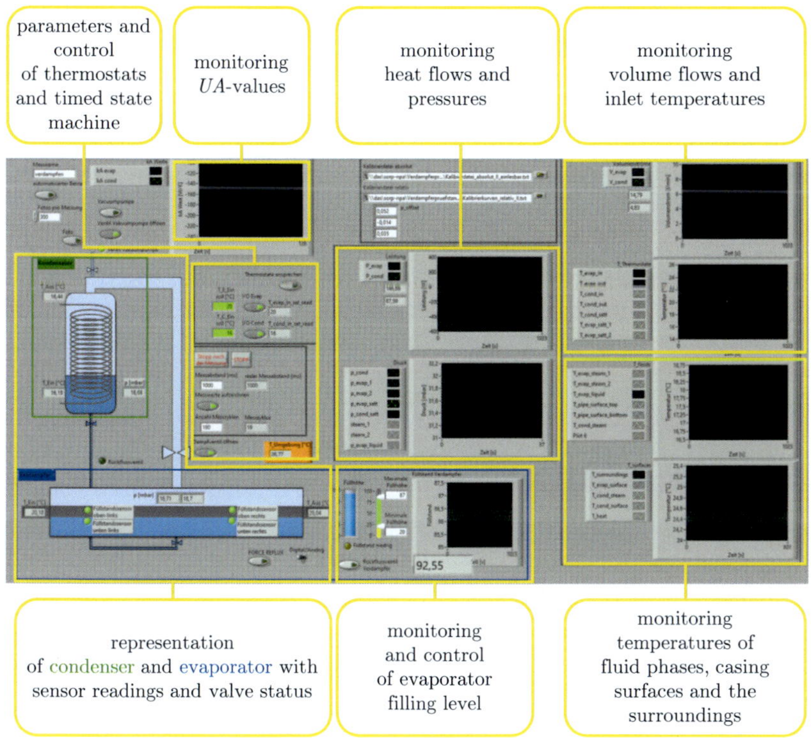

Figure B.9: Evaporation experiments: screenshot of the front panel of the LabVIEW program implemented to operate the experimental setup.

A second LabVIEW program was implemented to automate the experimental setup by successively starting experiments with predefined operation conditions. Thereby, multiple repetitions and screening of operational conditions were rendered possible without extensive staff support. A screenshot of the front panel of this LabVIEW program is shown in Figure B.10.

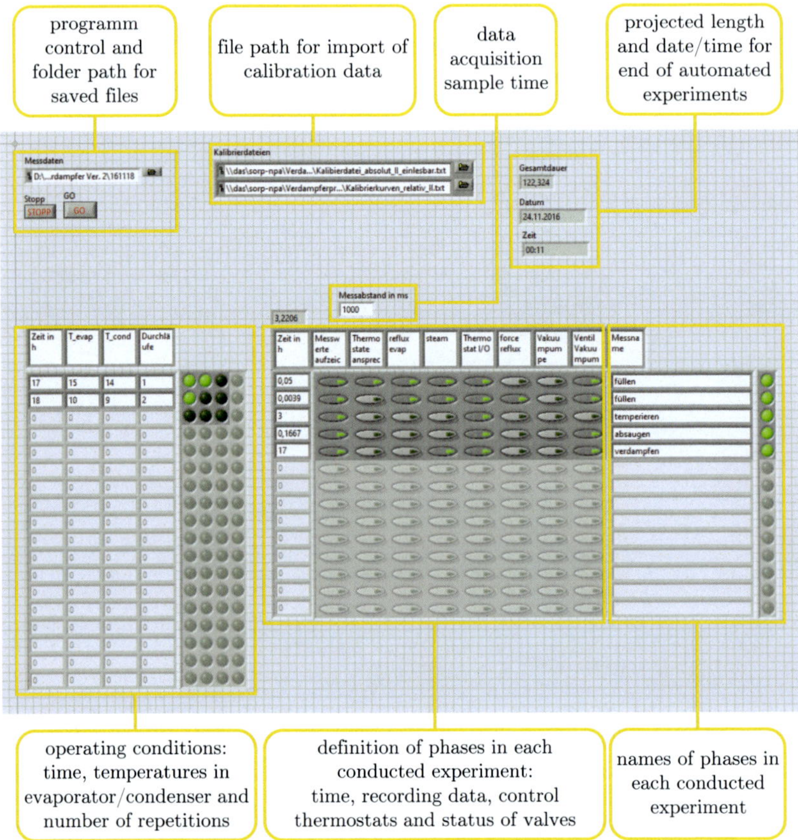

Figure B.10: Evaporation experiments: screenshot of the front panel of the LabVIEW program implemented to automate the operation of the experimental setup.

B.3 Equipment used in both experimental setups

Both data acquisition from sensors and control of valves status as well as the pumps and digital camera were realised by means of direct USB interfaces or analog-to-digital resp. digital-to-analog converters (technical details of employed hardware is summarised in Table B.10). The data acquisition hardware was connected to a standard computer by serial buses. Communication between the computer and the data acquisition hardware was realised with the software LabVIEW.

Table B.10: Experimental equipment: data acquisition used in experimental setups connected to a computer by serial bus and evaluated at a sampling rate of 1 Hz. The internal sampling rates of the modules differ, yet internal processing is used to only transfer the average of the measured values during the last second.

Data type per module	**Manufacturer**	**Model**	**Measured/actuated quantity**
1x analog in/out 1x digital in/out	Gantner Instruments	e.bloxx A1	universal
4x analog in 1x digital in/out	Gantner Instruments	e.bloxx A3	$0 - 10\,\mathrm{V}$ $4 - 20\,\mathrm{mA}$ (shunt)
4x analog in 1x digital in/out	Gantner Instruments	e.bloxx A4TC	thermocouples
2x analog in 1x digital in/out	Gantner Instruments	e.bloxx A5	Pt100 with 4-wire technique
4x analog out 1x digital in/out	Gantner Instruments	e.bloxx A9	$0 - 10\,\mathrm{V}$ $4 - 20\,\mathrm{mA}$
24x digital in/out	National Instruments	USB-6051	digital output
Multiple analog/ digital in/out	msl	DataScan 7000 series	universal

Details of the employed vacuum pumps are given in Table B.11. The vacuum pumps by the manufacturer Leybold GmbH are comparably large and powerful. However, they also dissipate large amounts of heat and increase the ambient temperature in the lab. Additionally, these vacuum pumps should only be operated for longer periods of time which is why they were only used if smaller pumps were not sufficient

Table B.11: Experimental equipment: vacuum pumps used in experimental setups.

Vacuum pump type	**Manufacturer**	**Model**	**Experimental Setup**
Dual-stage rotary vane	Leybold GmbH	TRIVAC D65B	Adsorption chiller
Dual-stage rotary vane	Leybold GmbH	TRIVAC D40B	Adsorption chiller Evaporation experiments
Dual-stage rotary vane	Rothenberger Werkzeuge GmbH	ROAIRVAC 1.5	Evaporation experiments

Appendix C

Analysis of measurement uncertainty

The commonly accepted framework to assess measurement uncertainty in standard technical measurement tasks is called "Guide to the expression of uncertainty in measurement" (GUM) (Joint Committee for Guides in Metrology, 2008, 2009). The GUM describes how to (I) estimate standard uncertainties for measurements, (II) combine the standard uncertainties of different measurements and (III) convert combined uncertainties to adjust the probability that the true value lies within the reported interval (e. g. for a standard uncertainty, approximately 68 % of the measurements lie within the stated interval; however, since the real probability distributions in measurement are usually not precisely described by the used mathematical statistical concepts, exact mathematical determination of this probability is impossible).

The GUM distinguishes two ways to estimate standard uncertainties: type A and type B evaluations:

Type A evaluation

Type A evaluation uses a statistical evaluation of n identically repeated measurements. The n measured values x_i are summarised by the arithmetic mean $\bar{x}$:

$$\bar{x} = \frac{1}{n}\sum_{i=1}^{n} x_i. \tag{C.1}$$

The experimental standard deviation s_x of the n measured values x_i is given by

$$s_x = \sqrt{\frac{1}{n-1}\sum_{i}^{n} (x_i - \bar{x})^2}. \tag{C.2}$$

According to GUM, the estimated standard uncertainty $u_{x,\mathrm{GUM}}$ of the variable x is then given by the experimental standard deviation of the mean $\bar{x}$

$$u_{x,\mathrm{GUM}} = s_{\bar{x}} = \frac{s_x}{\sqrt{n}}. \tag{C.3}$$

However, the number of repetitions n "should be large enough" (Joint Committee for Guides in Metrology, 2008), e. g. $n \geq 10$ (Birch, 2003). Birch (2003) suggests that the standard uncertainty for a single measurement, for which the known experimental standard deviation was derived from a set of prior experiments with $n < 10$, should be determined differently: here the estimated standard uncertainty u_x should be calculated from the experimental standard deviation s_x of the prior experiments including the according Student t-factor:

$$u_x = t_P s_x. \tag{C.4}$$

The Student t-factor t_P enlarges the uncertainty to compensate for insufficient repetitions n of the experiment. The Student t-factor is given in the GUM (Table G.2 in (Joint Committee for Guides in Metrology, 2008)) and depends on the degrees of freedom $\nu = n - 1$ and the probability P that the true value lies within the uncertainty interval.

Type B evaluation

Type B evaluation is employed for all quantities which have not been obtained from repeated measurements. Instead, information from manufacturer's specifications, certificates, previous measurements, general knowledge or other sources are used to evaluate the uncertainty of the quantity. If intervals with according uncertainties are given, they need to be converted to a standard uncertainty. Often, manufacturer's specifications state an interval, in which the measured value should lie, yet they very rarely specify the probability distribution of the stated interval. If information on the probability distribution is missing, the GUM suggests to assume a rectangular distribution, i. e. all expected values lie within the given interval and the probability is evenly distributed. A rectangular distribution with the interval $\pm a$ can be converted to a standard measurement uncertainty u according to:

$$u = \frac{a}{\sqrt{3}}. \tag{C.5}$$

For many technical measurement tasks, measured values are subject to data acquisition and further data reduction. The impact of data acquisition and reduction on the uncertainty of the final reported value is discussed in the next section.

Sensitivity analysis: propagation of uncertainties in data acquisition and reduction

To determine the uncertainty of calculated variables, the uncertainties can be propagated through all steps of data acquisition and reduction. The propagation of uncertainty through a calculation can be seen as a sensitivity analysis: to what extent does the output of a function change, if the input is altered? For functions of which a mathematical formulation $y = f(x_i)$ and its partial derivatives $\frac{\partial f}{\partial x_i}$ are known, the combined standard uncertainty $u_{c,f(x_i)}$ of the function $f(x_i)$ can be estimated based on a first-order Taylor series approximation of the function $f(x_i)$. Equation (C.6) is called the "law of propagation of uncertainty" (Joint Committee for Guides in Metrology, 2008) for uncorrelated input quantities

$$u_{c,f(x_i)} = \sqrt{\sum_i \left(\frac{\partial f}{\partial x_i}\right)^2 u_{x_i}^2}, \tag{C.6}$$

where u_{x_i} is the standard measurement uncertainty of the input x_i. For functions of which the mathematical formulation is not available (e. g. physical properties from tables), numerical approximations of the derivatives can be used instead. Thus, calculation routines for data reduction, which convert measured values to the final reported variables, should also incorporate propagation of the respective uncertainties in each calculation step.

Besides the uncertainties of the measurement sensors, further uncertainty can be added during data acquisition and reduction. In case the sensor's readings are collected by data acquisition hardware, the uncertainties of the used hardware and interface, e. g. $0 - 10\,\mathrm{V}$ or $4 - 20\,\mathrm{mA}$, also need to be considered. For some high-resolution sensors, e. g. high-quality pressure transducers and flowmeters, the contribution to uncertainty of the interface can be of the same order of magnitude as the uncertainty of the sensor itself. Often, state-of-the-art sensors internally employ digital sampling and modulate an analogous signal, e. g. $0 - 10\,\mathrm{V}$ or $4 - 20\,\mathrm{mA}$, for communication via the analogous interface with the data acquisition hardware. Thus, there are two consecutive transformations of the signal: from analogous to digital, again to analogous and back to digital. To reduce the overall uncertainty, the unnecessary second transformation can be avoided by using sensors that offer digital interfaces to obtain the readings. Hence, wherever possible, digitally connected sensors should be used. Uncertainties of all other inputs, which are introduced during data reduction, e. g. physical properties, parameters, etc., can and should also be considered in Equation (C.6). Thus, the whole measurement chain from the sensor to data acquisition and through data reduction can be included in a thorough analysis of measurement uncertainty.

The combination of different contributions to the combined measurement uncertainty in Equation (C.6) strictly requires the use of standard uncertainties as input. For the final variable, the reported interval, in which the true value of the variable should lie, can be expanded by a coverage factor k. The coverage factor k is multiplied with the standard combined uncertainty u_c to obtain an expanded uncertainty, thereby increasing the encompassed part of the probability distribution and, thus, the probability that the true value lies within the reported interval. For example, according to the Joint Committee for Guides in Metrology (2008) a coverage factor of $k = 2$ ($k = 3$) produces an interval having a level of confidence of approximately 95 % (99 %). In this thesis, all uncertainties are given with a coverage factor $k = 1$ leading to reported intervals having a level of confidence of approximately 68 %.

Appendix D

Supplementary Information to Chapter 4

In this appendix, Supplementary Information to chapter 4 "Validating the setup and procedure for the evaporation experiments" is given. For the compared experiments between RWTH and ISE, the SI includes the time resolved input conditions (Appendix D.1), visualisation of the uncertainty in measurement (Appendix D.2), discussion of other potential reasons for the observed deviations between the measurements at RWTH and ISE (Appendix D.3), the time-resolved relative filling level for the experiments at RWTH (Appendix D.4) and a discussion of contact angle determination (Appendix D.5).

Contents of this appendix have been reprinted from:

J. Seiler, R. Volmer, D. Krakau, J. Pöhls, F. Ossenkopp, L. Schnabel and A. Bardow. "Capillary-assisted evaporation of water from finned tubes – Impacts of experimental setups and dynamics". *Applied Thermal Engineering* 165 (2020), 114620, with permission from Elsevier. Contributions of the author: writing the draft as one of the principal authors, adapting the design and building of the experimental setup at RWTH, planning the measurement procedure, supervising the experiments at RWTH, implementing the data reduction, and evaluating the results.

Reprinted passages use the *pluralis modestiae* "we" as an alternative to the passive voice.

D.1 Input conditions for experiments

In order to be able to compare the results of the experimental setups at RWTH and ISE, the input conditions of the experiments need to be as similar as possible. The input conditions (volume flow, inlet temperature and evaporator pressure) are given in Figure D.1 for parameter set lowDF (corresponding to Figure 4.3) and Figure D.2 for parameter set highDF (corresponding to Figure 4.4).

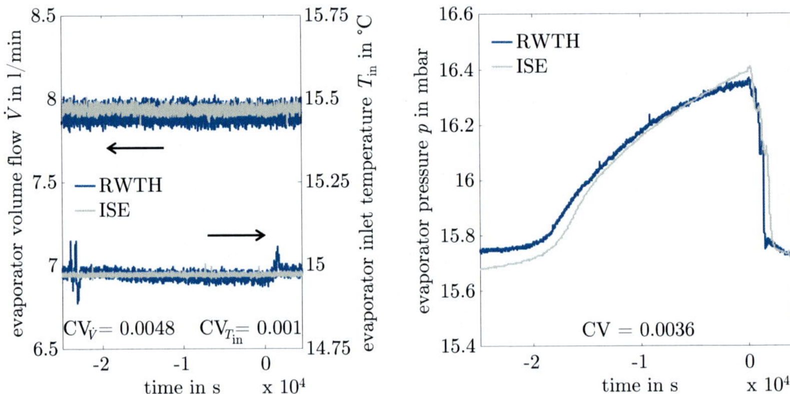

Figure D.1: Operating conditions for experiments with parameter set lowDF plotted over time. Left: evaporator volume flow $\dot{V}$ and inlet temperature T_{in}, right: evaporator pressure p. Both experiments (RWTH and ISE) are synchronized at the end of the experiment (time=0) when the first tube loses contact to the refrigerant pool.

During thin-film evaporation until the contact between the tubes and the water is lost, the input conditions of the presented experiments show good agreement within uncertainty of measurement in both experimental setups. This results in very low CV given in Figure D.1 and Figure D.2.

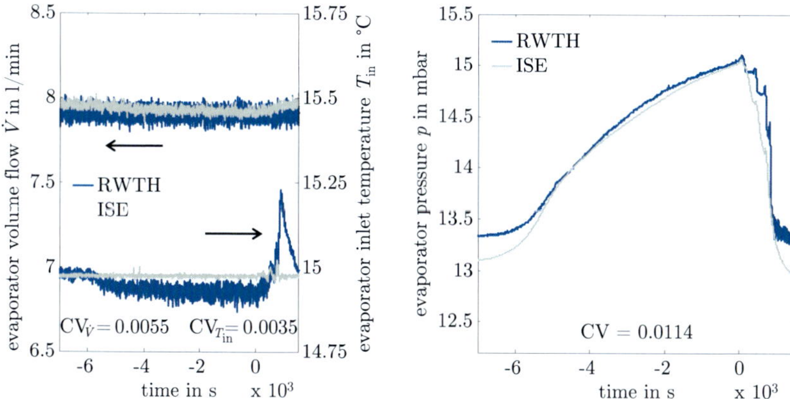

Figure D.2: Operating conditions for experiments with parameter set highDF plotted over time. Left: evaporator volume flow $\dot{V}$ and inlet temperature T_{in}, right: evaporator pressure p. Both experiments (RWTH and ISE) are synchronized at the end of the experiment (time=0) when the first tube loses contact to the refrigerant pool.

The evaporator volume flow and inlet temperature are constant and almost identical for both setups (left in Figure D.1 and Figure D.2). However, the inlet temperature of the RWTH setup shows peaks at the beginning and the end of the evaporation process for both parameter sets due to insufficient control of the thermostat. Nevertheless, these peaks do not affect the experiments as they do not occur during the relevant thin-film evaporation. Furthermore, the inlet temperature of the RWTH setup is slightly lower for the parameter set highDF once the thin-film evaporation started, since the thermostat was not able to keep the temperature constant. We repeated the experiment with slightly higher inlet temperatures in the RWTH setup and did not obtain considerable differences. Thus, the difference in inlet temperature can be neglected.

The measured evaporation pressure shows similar curve shapes and almost identical maximum and minimum values. Although the measured values are not identical, their values coincide within the uncertainty of measurement. The measured pressures are not constant during the experiment as they adapt to the current UA-values of the evaporator and condenser since the fluid inlet temperatures of both components are controlled at fixed values by the thermostats.

D.2 Uncertainty in measurement

The uncertainty of measurement for all calculated quantities plotted in Figure 4.3 and Figure 4.4 are given in Figure D.3 and Figure D.5. The uncertainties were calculated using Equation (C.6) and sensor uncertainties given in Table 3.3 and Table 4.2 according to the "Guide to the expression of uncertainty in measurement" (GUM) using a coverage factor of $k = 1$ (Joint Committee for Guides in Metrology, 2008). In Figure D.4 and Figure D.6, the uncertainty of measurement is plotted as background behind the actually measured values given in Figure 4.3 and Figure 4.4. The used colours in Figure D.4 and Figure D.6 differ from the colours in Figure 4.3 and Figure 4.4 to improve readability of the uncertainty range.

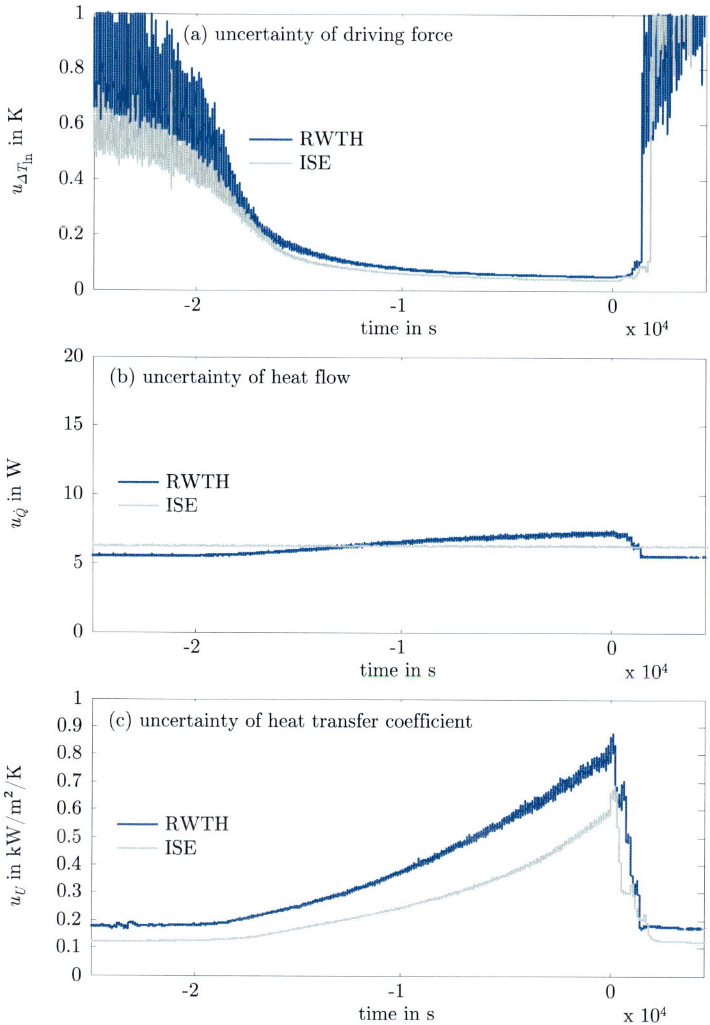

Figure D.3: Standard uncertainty in measurement for the quantities driving force ΔT_{ln} (a), heat flow $\dot{Q}$ (b) and overall heat transfer coefficient U (c) shown in Figure 4.3 for experiments with parameter set lowDF (Table 4.3) plotted over time. Both experiments (RWTH and ISE) are synchronized at the end of the experiment (time=0) when the first tube loses contact to the refrigerant pool.

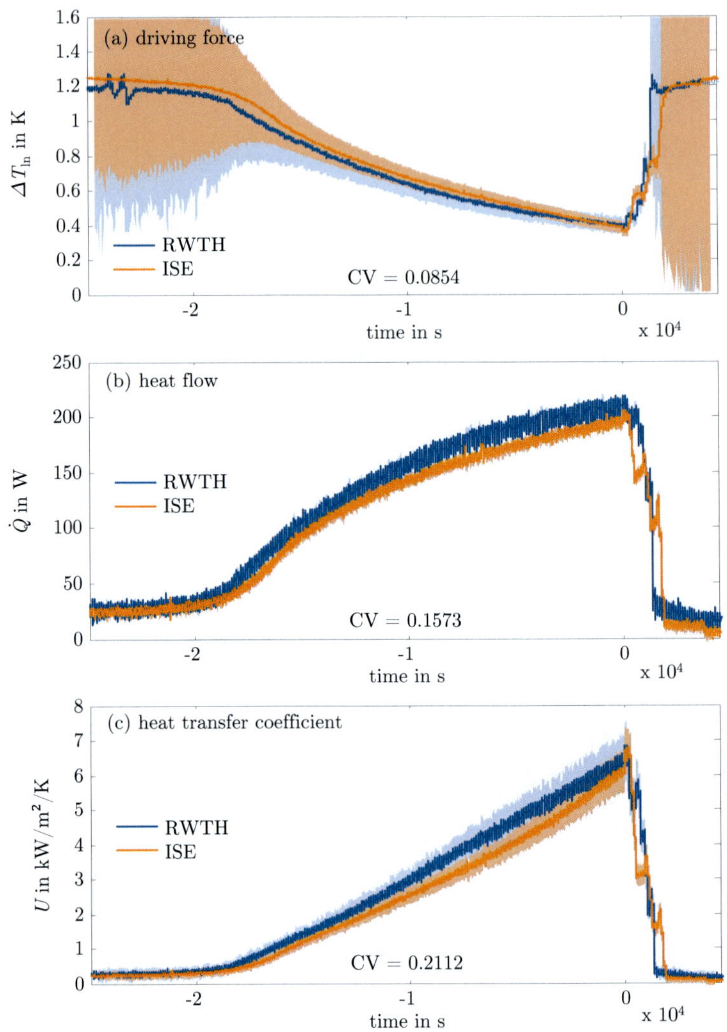
(a) driving force
ΔT_{ln} in K
RWTH
ISE
CV = 0.0854
time in s
x 10^4
(b) heat flow
$\dot{Q}$ in W
RWTH
ISE
CV = 0.1573
time in s
x 10^4
(c) heat transfer coefficient
U in kW/m^2/K
RWTH
ISE
CV = 0.2112
time in s
x 10^4

Figure D.4: (figure on previous page) Driving force ΔT_{ln} (a), heat flow $\dot{Q}$ (b) and overall heat transfer coefficient U (c) shown in Figure 4.3 together with uncertainties of measurement (shown in Figure D.3) in background for experiments with parameter set lowDF (Table 4.3) plotted over time. Both experiments (RWTH and ISE) are synchronized at the end of the experiment (time=0) when the first tube loses contact to the refrigerant pool. The coefficient of variation CV (Equation (4.1)) is given in each plot to quantify the deviation between the experiments at RWTH and ISE.

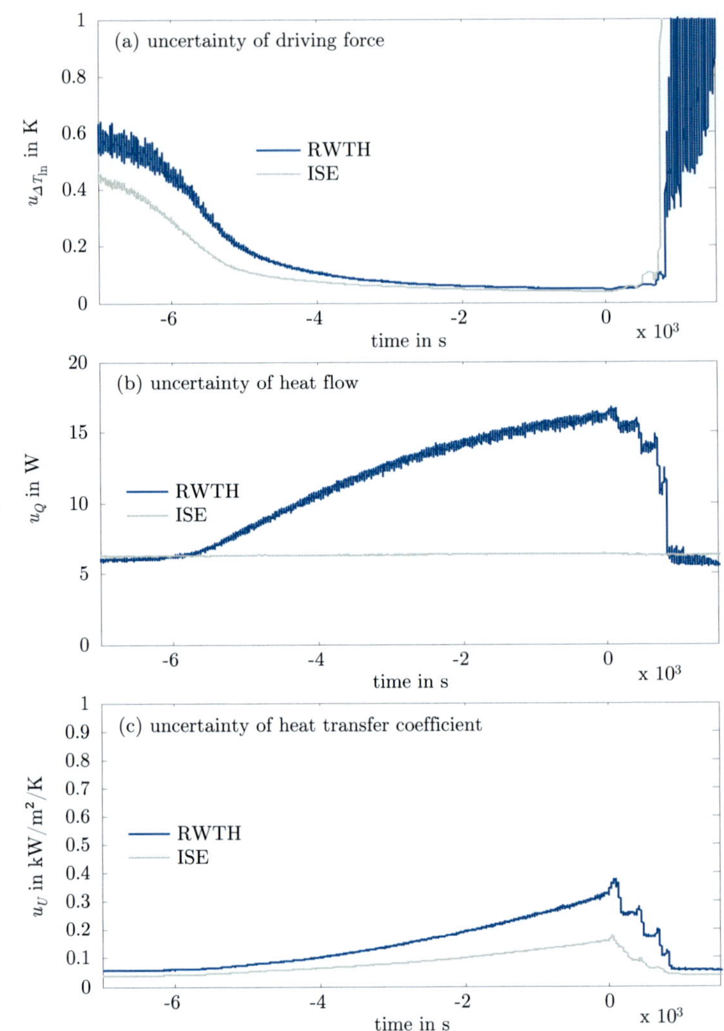
(a) uncertainty of driving force
$u_{\Delta T_{\ln}}$ in K
RWTH
ISE
time in s
x 10³
(b) uncertainty of heat flow
u_Q in W
RWTH
ISE
time in s
x 10³
(c) uncertainty of heat transfer coefficient
u_U in kW/m²/K
RWTH
ISE
time in s
x 10³

Figure D.5: (figure on previous page) Standard uncertainty in measurement for the quantities driving force ΔT_{ln} (a), heat flow $\dot{Q}$ (b) and overall heat transfer coefficient U (c) shown in Figure 4.4 for experiments with parameter set highDF (Table 4.3) plotted over time. Both experiments (RWTH and ISE) are synchronized at the end of the experiment (time=0) when the first tube loses contact to the refrigerant pool.

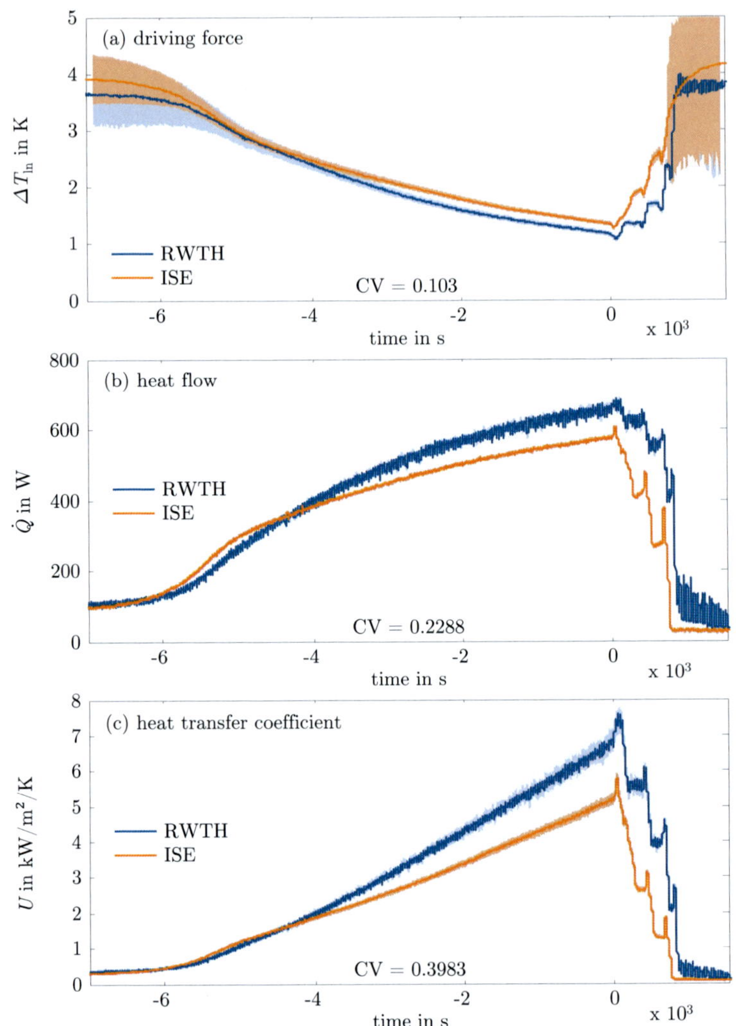

(a) driving force
ΔT_{ln} in K
RWTH
ISE
CV = 0.103
time in s
x 10³
(b) heat flow
$\dot{Q}$ in W
RWTH
ISE
CV = 0.2288
time in s
x 10³
(c) heat transfer coefficient
U in kW/m²/K
RWTH
ISE
CV = 0.3983
time in s
x 10³

Figure D.6: (figure on previous page) Driving force ΔT_{ln} (a), heat flow $\dot{Q}$ (b) and overall heat transfer coefficient U (c) shown in Figure 4.4 together with uncertainties of measurement (shown in Figure D.5) in background for experiments with parameter set highDF (Table 4.3) plotted over time. Both experiments (RWTH and ISE) are synchronized at the end of the experiment (time=0) when the first tube loses contact to the refrigerant pool. The coefficient of variation CV (Equation (4.1)) is given in each plot to quantify the deviation between the experiments at RWTH and ISE.

D.3 Discussion of other potential reasons for observed deviations between RWTH and ISE

The three aspects that we consider most likely to be responsible for the discrepancies in the measurement results from RWTH and ISE are discussed in detail in Section 4.3. Further factors which might generally influence evaporation measurements but which are considered negligible in our case are pointed out in detail below:

(IV) **Heat exchanger design.** While at RWTH evaporation can only take place at the finned tube sections, at ISE also connecting plain tube sections and the "U"-bends are in contact with refrigerant and can potentially contribute to evaporation. However, refrigerant which was condensed on the plain tube sections above the pool surface evaporated completely before the evaluated time interval. Thus, results cannot be affected by the additional evaporation source. The plain "U"-bends are immersed in the refrigerant pool and form contact lines, but the contact line length is very small compared to the contact line length of the finned tube section. Therefore, a noticeable impact on the measurement is not expected.

(V) **Tube alignment.** Ideally, the evaporator tubes should be aligned perfectly horizontally to achieve a homogeneous refrigerant distribution in the fins interstices and a synchronous detachment of the refrigerant meniscus at the end of the experiment. If the tubes are slightly tilted, altered and undefined wetting conditions could change evaporation power and U-values. The ISE setup is generally more susceptible to improper tube alignment since the heat exchanger is not mechanically fixed in its place except by the hose connections. However, measured U-values should show an extended period of time for detachment if the tubes were imprecisely aligned. As this period of time is very similar at RWTH and ISE, the precision of tube alignment is not considered as possible reason for the deviations in the presented measurements.

(VI) **Constructional conditions for vapour flow.** Number, geometry and arrangement of connection lines and valves between evaporator and condenser can induce a pressure drop and determine fluid flow dynamics. Since the system pressure is measured directly inside the evaporator chamber, pressure drops in the connection to the condenser cannot distort the results in the evaporator. The vacuum chambers have a rather large and open volume so that a homogeneous pressure distribution within them is assumed. Vapour flow dynamics inside the evaporator chamber are considered to be negligible for the pressure measurement since applied driving forces and thereby vapour velocities at the location of the pressure transducers are small.

(VII) **Pressure difference.** A process-related factor, which should be considered in the interpretation of the results, is the slight difference in pressure between RWTH and ISE measurements (see Figure D.1 and Figure D.2). Since neither of the two setups can directly control pressure, the pressure in the evaporator chamber is rather an output quantity which dynamically adapts to fluid temperatures, volume flows and heat transfer capabilities of the evaporator and condenser. The resulting maximum pressure difference occurred in the flooded period and was approx. 0.1 mbar for parameter set lowDF and approx. 0.3 mbar for parameter set highDF. During the second half of the partially flooded period – in which the largest differences of U and $\dot{Q}$ between RWTH and ISE results appear – the pressure difference was only approx. 0.05 mbar (parameter set lowDF) and 0.1 mbar (parameter set highDF). This rather small difference is within the uncertainty of measurement of the pressure transducers and therefore, we consider it as not sufficient for having a remarkable effect on the refrigerant's material properties and, thus, do not take this factor into account as a reason for the measured differences.

(VIII) **Non-condensable gases.** Since the calculation of saturation temperature from the absolute pressure signal is based on the assumption of a pure water atmosphere, the presence of non-condensable gases inside the evaporator chamber would distort this calculation. Generally, a certain leakage of gaseous compounds from the environment to the interior cannot be avoided with any vacuum system and must be expected in both setups. However, non-condensable gases can enter the vacuum also in a different way: our observations indicate that during exposure to air, gaseous components are adsorbed on the heat exchanger surface and/or dissolved in the refrigerant. After exposition to vacuum, the gaseous components are not directly released which can lead to significant deviations of the first measurements after installation compared to the subsequent ones. Consequently, measurement results are only dependable if a series of reproduction measurements do not show any significant changes any more. To control leakage of non-condensable gases inside the vacuum chamber, evacuation procedures are necessary. While at RWTH non-condensables were constantly withdrawn from the system, evacuation at ISE only took place before the refrigerant was condensed on the evaporator. Thus, it can be assumed that the ISE evacuation procedure leads to higher partial pressures of non-condensable gases than the RWTH procedure. However, quantification of the partial pressures is not possible at either of the two setups. Besides quantity, the distribution of non-condensable gases within the closed system is an important issue: supposing that the non-condensable gases were equally distributed within the system, they would massively falsify the calculated saturation temperature. However, in a closed evaporator-condenser system, non-condensable gases usually accumulate at the place of condensation (van Carey, 2018). Therefore, it can be assumed that only a minor fraction of the non-condensable gases remained in the evaporator and distorted the pressure signal. This statement is supported by additional measurements at ISE with intermittent evacuation phases during evaporation which did not yield considerably different results in the heat transfer coefficient. Despite the lack of knowledge of the exact amount of non-condensable gases inside the evaporators, this factor can definitely not be the cause for the difference in $\bar{U}$-values between RWTH and ISE: if non-condensable gases distorted the pressure signal at ISE, the corrected saturation temperature would be lower than the evaluated one which would – according to Equations 3.3 and 3.6 – result in a higher driving force and a lower U-value. Since the calculated U-values of the ISE measurements are already lower than of RWTH, a correction for non-condensable gases would rather increase the differences than diminish them. However, the presence of non-condensable gases can affect the conclusions drawn on the impact of the driving force: given that the measurement duration of parameter set lowDF was about four times as long as of parameter set highDF, a considerably greater amount of non-condensable gases can be assumed for parameter set lowDF for the ISE measurements. Consequently – if non-condensables play a relevant role at all – a reduction of U-values would affect parameter set lowDF much stronger than parameter set highDF. The two U curves from ISE would then come closer together and the concluded effect of driving force on heat transfer coefficient would lessen.

D.4 Relative filling level f_{rel} detected optically in experimental setup at RWTH

The optically detected relative filling levels f_{rel} in the experimental setup at RWTH are given in Figure D.7 for parameter set lowDF and in Figure D.8 for parameter set highDF. Details on the optical detection system can be found in Figure 3.6 (c) and Section 3.2.2.

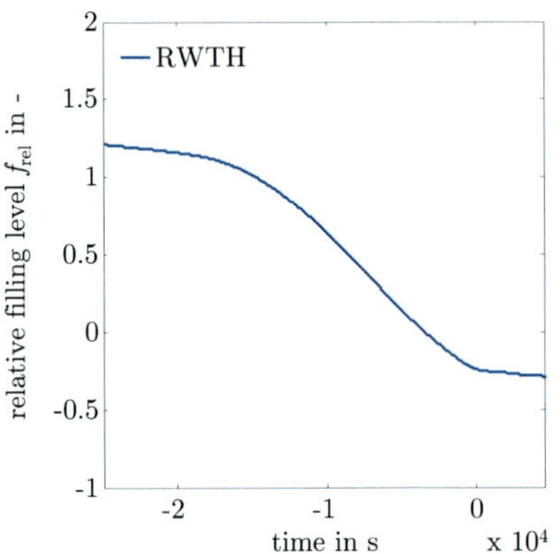

Figure D.7: Relative filling level f_{rel} of the experiment at RWTH with parameter set lowDF shown in Figure 4.3 plotted over time.

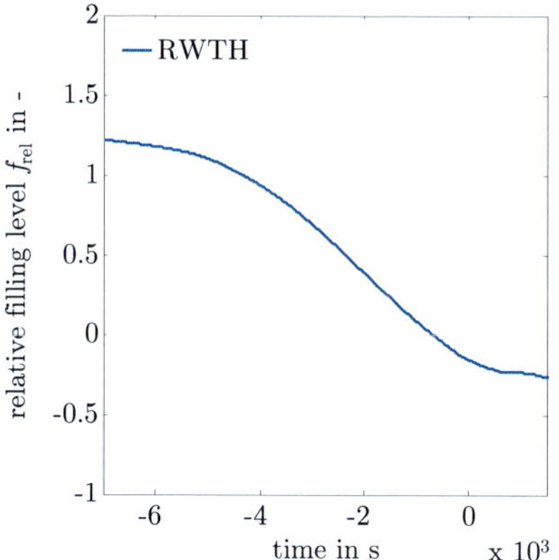

Figure D.8: Relative filling level f_{rel} of the experiment at RWTH with parameter set highDF shown in Figure 4.4 plotted over time.

D.5 Contact angle

Unfortunately, we were unable to determine the wetting angle on the surface of the heat exchanger's fins (material Cu, C12200) at the process conditions of the experiments. Evaluating pictures taken from the side of the experimental setup at RWTH (cf. Figure 4.1) gives a hint on contact angle and indicates that it changes at different process conditions, yet the quality and resolution of the pictures impede proper quantitative assessment. The height of the water meniscus between the fins at process conditions is also difficult to determine through the existing windows. The space between the fins is usually almost completely filled with refrigerant, although the picture in Figure 4.1 reveals differences in the height of the meniscus between different fins. Zooming in on the picture in Figure 4.1 shows that on the upper end of the tube´s fins, the distance between the highest point of refrigerant and the lowest point of the refrigerant in the middle between two fins is usually smaller than the fin width (0.15 mm) in the picture. However, evaluation of distances in this picture can be prone to large errors due to resolution and sharpness of the picture as well as slight perspective deviation.

Appendix E

Supplementary Information to Chapter 6

In this appendix, Supplementary Information to chapter 6 "Water-based adsorption cooling below 0 °C" is given. For all conducted experiments, further temperature and pressure readings are presented in the following.

Contents of Appendix E have been reprinted from:

J. Seiler, J. Hackmann, F. Lanzerath, and A. Bardow. "Refrigeration below zero °C: Adsorption chillers using water with ethylene glycol as antifreeze". *International Journal of Refrigeration* 77 (2017), pp. 39–47, with permission from Elsevier. Contributions of the author: writing the draft as principal author, adapting the design and building of the experimental setup, planning the measurement procedure, supervising the experiments, implementing the data reduction, and evaluating the results.

Reprinted passages use the *pluralis modestiae* "we" as an alternative to the passive voice.

Further data from the used temperature sensors and pressure transducers are given in Figures E.1 to E.5.

The plotted temperatures include in- and outlet temperatures as shown in Figure 6.4 (b). Additionally, measured surface temperatures of the component's metal casings are included in the graphs. The casing temperatures were measured using thermocouple sensors, thus, their uncertainty is higher than the given uncertainty of the in- and outlet temperatures. Furthermore, a local temperature of the liquid phase in the evaporator is shown. It is important to note that the temperature of the liquid phase is not uniform in the evaporator. Thus, the temperature of the liquid phase may differ in the evaporator at different locations. Pressure readings are also included in the figures. However, the used pressure transducers have a large uncertainty according the manufacturer (technical datasheet: 10 mbar). Thus, we prefer to use the pressure transducers only to operate the test stand and not for scientific measurements even though the perceived uncertainty seems much smaller.

In Figure E.1, supplementary data for Figure 6.5 is given. Evaporation at 0 °C for 30 % and 60 % glycol mass fraction.

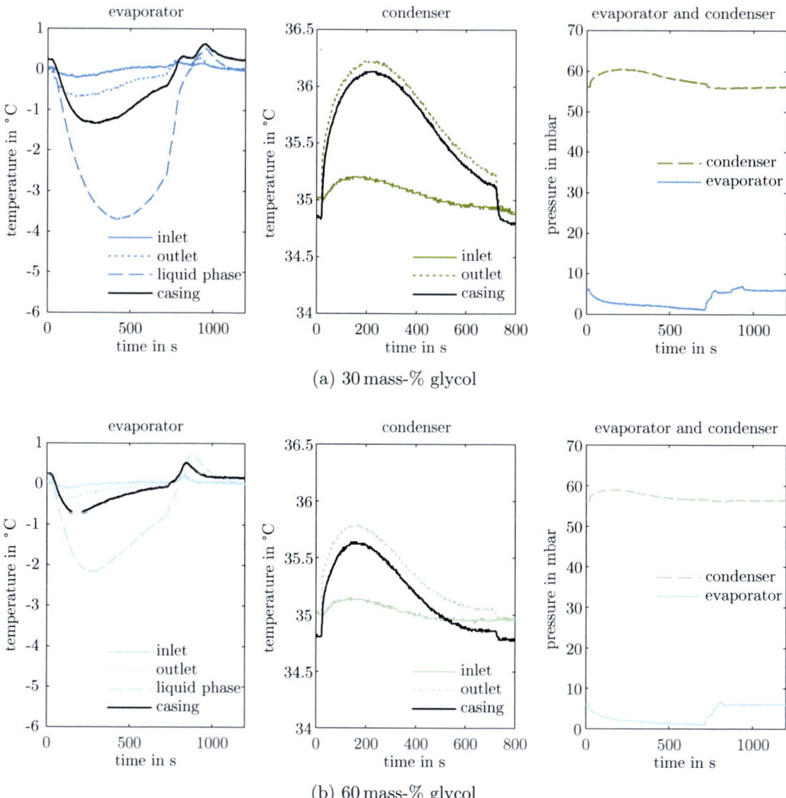

(a) 30 mass-% glycol

(b) 60 mass-% glycol

Figure E.1: Temperatures and pressure in evaporator (during adsorption phase) and condenser (during desorption phase) for experiments at 0 °C evaporation temperature for 30 mass-% glycol (a) and 60 mass-% glycol (b). Note that pressure curves are plotted in one diagram for evaporator and condenser although they have been measured in consecutive phases.

In Figure E.4, supplementary data for Figure 6.6 is given. Evaporation at 5 °C for 0 %, 30 % and 60 % glycol mass fraction.

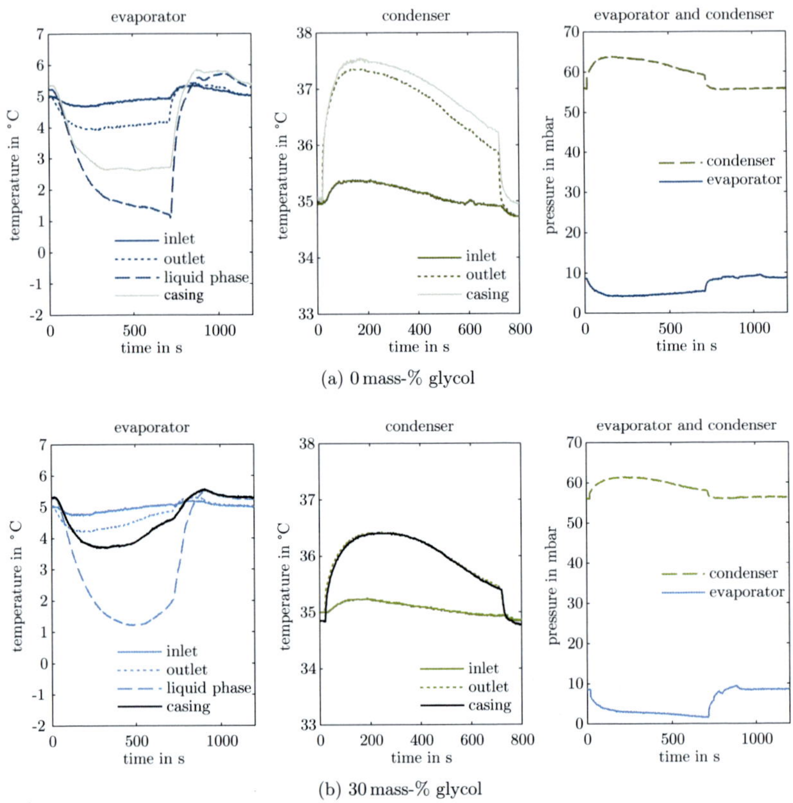

(a) 0 mass-% glycol

(b) 30 mass-% glycol

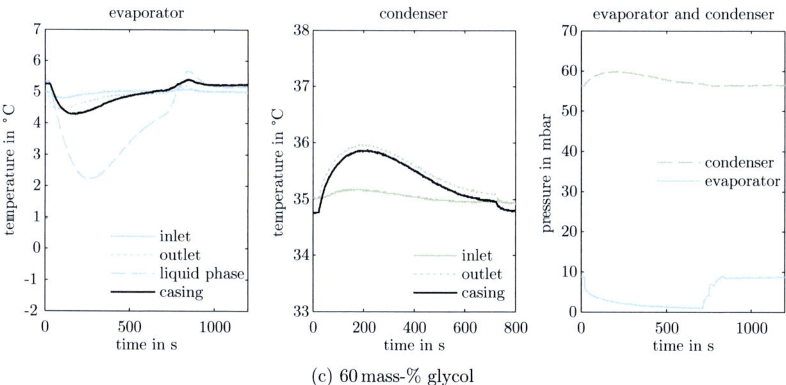

(c) 60 mass-% glycol

Figure E.4: (figure on previous page and continued on this page) Temperatures and pressure in evaporator (during adsorption phase) and condenser (during desorption phase) for experiments at 5 °C evaporation temperature for 0 mass-% glycol (a), 30 mass-% glycol (b) and 60 mass-% glycol (c). Note that pressure curves are plotted in one diagram for evaporator and condenser although they have been measured in consecutive phases.

In Figure E.5, supplementary data for Figure 6.7 is given. Evaporation at 10 °C for 0 % and 30 % glycol mass fraction for cycle times reaching equilibrium.

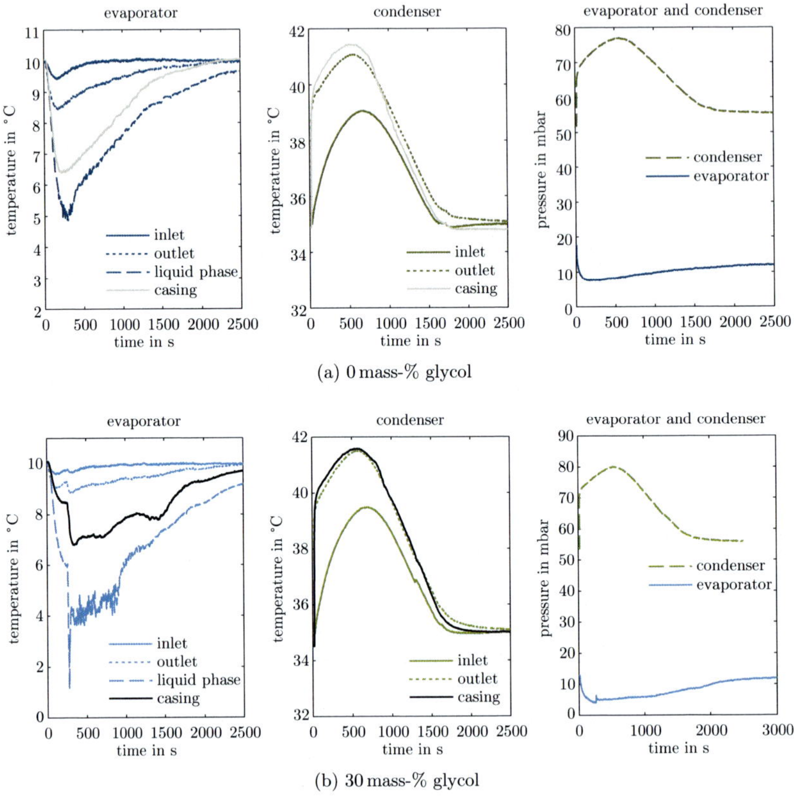

Figure E.5: Temperatures and pressure in evaporator (during adsorption phase) and condenser (during desorption phase) for experiments at 10 °C evaporation temperature and cycle times reaching equilibrium for 0 mass-% glycol (a) and 30 mass-% glycol (b). Note that pressure curves are plotted in one diagram for evaporator and condenser although they have been measured in consecutive phases.

Appendix F

Publications and student theses

Links to most of the cited references are available in the digital version of this document and have been last accessed on October 2nd, 2020.

List of publications

Journal papers

Seiler, J., Volmer, R., Krakau, D., Pöhls, J., Ossenkopp, F., Schnabel, L., and Bardow, A. (2020). Capillary-assisted evaporation of water from finned tubes – impacts of experimental setups and dynamics. *Applied Thermal Engineering*, 165:114620.

Bau, U., Baumgärtner, N., Seiler, J., Lanzerath, F., Kirches, C., and Bardow, A. (2019). Optimal operation of adsorption chillers: first implementation and experimental evaluation of a nonlinear model-predictive-control strategy. *Applied Thermal Engineering*, 149:1503–1521.

Seiler, J., Lanzerath, F., Jansen, C., and Bardow, A. (2019). Only a wet tube is a good tube: understanding capillary-assisted thin-film evaporation of water for adsorption chillers. *Applied Thermal Engineering*, 147:571–578.

Gibelhaus, A., Tangkrachang, T., Bau, U., Seiler, J., and Bardow, A. (2019). Integrated design and control of full sorption chiller systems. *Energy*, 185:409–422.

Seiler, J., Hackmann, J., Lanzerath, F., and Bardow, A. (2017). Refrigeration below zero °C: adsorption chillers using water with ethylene glycol as antifreeze. *International Journal of Refrigeration*, 77:39–47.

Lanzerath, F., Seiler, J., Erdogan, M., Schreiber, H., Steinhilber, M., and Bardow, A. (2016). The impact of filling level resolved: Capillary-assisted evaporation of water for adsorption heat pumps. *Applied Thermal Engineering*, 102:513–519.

Lanzerath, F., Bau, U., Seiler, J., and Bardow, A. (2015). Optimal design of adsorption chillers based on a validated dynamic object-oriented model. *Science and Technology for the Built Environment*, 21(3):248–257.

Conference contributions

Entrup, M., Kappelhoff, C., Seiler, J., and Bardow, A. (2020). Wasser/Ethanol – Adsorptionskältemaschine unter 0 °C (in German). *Tagung Deutscher Kälte- und Klimatechnischer Verein e. V. (DKV 2020)*, Magdeburg, Germany (accepted).

Seiler, J., Volmer, R., Krakau, D., Pöhls, J., Ossenkopp, F., Schnabel, L., and Bardow, A. (2020). Capillary-assisted evaporation of water from finned tubes: impacts of dynamics and experimental setups. *International Sorption Heat Pump Conference (ISHPC 2020)*, Berlin, Germany (accepted).

Henninger, M., Gilges, M., Rustam, L., Ernst, S.-J., Velte, A., Seiler, J., and Bardow, A. (2020). Heat and mass transfer kinetics of MOF-coatings from Al-fumarate and CAU-10(Al)-H from IR-LTJ experiments: the impact of characteristic times. *International Sorption Heat Pump Conference (ISHPC 2020)*, Berlin, Germany (accepted).

Engelpracht, M., Yang, Z., Gluesenkamp, K., Turnaoglu, T., Seiler, J., and Bardow, A. (2020). SorpPropLib: an open-source database for sorption equilibrium properties. *International Sorption Heat Pump Conference (ISHPC 2020)*, Berlin, Germany (accepted).

Postweiler, P., Gibelhaus, A., Engelpracht, M., Seiler, J., and Bardow, A. (2020). Experimental validation of a dynamic adsorption chiller model using optimal experimental design. *International Sorption Heat Pump Conference (ISHPC 2020)*, Berlin, Germany (accepted).

Gibelhaus, A., Postweiler, P., Seiler, J., and Bardow, A. (2020). Computationally efficient, experimentally validated adsorption chiller model using a plug-flow-based modelling approach. *International Sorption Heat Pump Conference (ISHPC 2020)*, Berlin, Germany (accepted).

Entrup, M., Kappelhoff, C., Seiler, J., and Bardow, A. (2020). Water and ethanol as refrigerant mixture: towards adsorption cooling below 0 °C. *International Sorption Heat Pump Conference (ISHPC 2020)*, Berlin, Germany (accepted).

Engelpracht, M., Driesel, J., Jürgens, B., Seiler, J., and Bardow, A. (2020). Transforming heat from 90 °C to 110 °C: demonstration of a lab-scale adsorption heat transformer (AdHT). *International Sorption Heat Pump Conference (ISHPC 2020)*, Berlin, Germany (accepted).

Seiler, J., Volmer, R., Krakau, D., Pöhls, J., Ossenkopp, F., Schnabel, L., and Bardow, A. (2020). Kapillarunterstützte Verdampfung von Wasser an Rippenrohren: Einfluss von Messaufbau und Dynamik (in German). In *Jahrestreffen der ProcessNet-Fachgruppe Wärme- und Stoffübertragung (WSUE 2020)*, Erfurt, Germany.

Entrup, M., Kappelhoff, C., Seiler, J., and Bardow, A. (2019). Wasser und Ethanol als Kältemittelgemisch: Ein neuer Zugang zu Adsorptionskältemaschinen unter 0 °C (in German). In *Thermodynamik-Kolloquium*, Duisburg, Germany.

Engelpracht, M., Seiler, J., and Bardow, A. (2019). Warmer Kaffee aus kaltem Kaffee: Experimenteller Nachweis der Adsorptionswärmetransformation (in German). In *Thermodynamik-Kolloquium*, Duisburg, Germany.

Seiler, J., Lanzerath, F., Jansen, C., and Bardow, A. (2018). Get your tubes wet: capillary-assisted thin-film evaporation of water for adsorption chillers. In Riehl, R. R., Preißinger, M., Eames, I. W. and Tierney, M., editors, *Proceedings of the 8th Heat Powered Cycles Conference (HPC 2018), Book of Abstracts*, Bayreuth, Germany.

Seiler, J., Jörres, S., Lanzerath, F., and Bardow, A. (2016). Thin-film evaporation of the natural refrigerant water from copper tubes with porous coatings at low pressures. In *IVth International Symposium on Innovative Materials for Processes in Energy Systems (IMPRES 2016)*, Taormina, Italy.

Seiler, J., Jörres, S., Lanzerath, F., and Bardow, A. (2016). Verdampfung des natürlichen Kältemittels Wasser an porösen Oberflächen (in German). In *Jahrestreffen der ProcessNet-Fachgruppe Wärme- und Stoffübertragung (WSUE 2016)*, Kassel, Germany.

Seiler, J., Hackmann, J., Lanzerath, F., and Bardow, A. (2015). Adsorptionsgestützte Kühlung unter 0 °C mit Ethylenglykol-Wasser-Gemischen (in German). In *Thermo-*

dynamik-Kolloquium, Bochum, Germany.

Seiler, J., Hackmann, J., Lanzerath, F., and Bardow, A. (2015). Adsorption-based low temperature refrigeration using water–ethylene glycol mixtures. In *Proceedings of the 24th IIR International Congress of Refrigeration (ICR 2015)*, Yokohama, Japan, International Institute of Refrigeration, Paris, France.

Lanzerath, F., Seiler, J., Bau, U., and Bardow, A. (2014). A modular experimental and simulation approach for the systematic development of adsorption heat pumps. In Radermacher, R., editor, *International Sorption Heat Pump Conference (ISHPC 2014)*, Maryland, United States, pages 117–126. Curran, Red Hook, NY.

Student theses supervised during this work

Cottmann, B. (2019). Automatisierung einer Füllstandssteuerung und Implementierung einer Druckregelung für einen kapillargestützten Dünnfilmverdampfer (in German). Bachelor's thesis, RWTH Aachen University, Germany.

Kastner, D. (2018). Implementierung eines 0-D Divertor Modells und Weiterentwicklung eines Tokamak-Systemcodes für die Durchführung von Parameterstudien zur Leistungsabfuhr in einem Kernfusionsreaktor (in German). Master's thesis, RWTH Aachen University, Germany.

Derwein, D. (2017). Experimentelle Untersuchung von Wasser-Ethanol Arbeitsmittelgemischen in Adsorptionswärmepumpen für Kältebereitstellung unterhalb von 0 °C (in German). Bachelor's thesis, RWTH Aachen University, Germany.

Krakau, D. (2017). Charakterisierung strukturierter Hochleistungsverdampferrohre zur kapillargestützten Dünnfilmverdampfung von Wasser (in German). Bachelor's thesis, RWTH Aachen University, Germany.

Göbel, S. (2016). Simulation und experimentelle Untersuchung der Dünnfilmverdampfung an porös beschichteten Hochleistungsverdampferrohren in einer Adsorptionskältemaschine (in German). Bachelor's thesis, RWTH Aachen University, Germany.

Kemetmüller, D. (2016). Simulative und experimentelle Untersuchung porös beschichteter Verdampferrohre für Adsorptionskältemaschinen (in German). Master's thesis, RWTH Aachen University, Germany.

Baumgärtner, N. (2016). Implementation and experimental validation of a model predictive control for adsorption heat pumps. Master's thesis, RWTH Aachen University.

Bohn, T. (2016). Untersuchung eines Membrandesorbers: Auslegung, Konstruktion, Aufbau und Experiment (in German). Bachelor's thesis, RWTH Aachen University, Germany.

Johansen, A. (2016). Simulation of absorption chillers using ionic liquids and deep eutectic solvents as working fluids. Pre-master specialization project, RWTH Aachen University, Germany.

Fröde, F., Vogelsang, M., and Zibunas, C. (2015). Experimentelle und simulative Untersuchung von Glykol-Wasser-Gemischen als Arbeitsmittel in einer Adsorptionswärmepumpe (in German). Project work, RWTH Aachen University, Germany.

Granacher, J. (2015). Analysis of a thermo control device for water. Project work, RWTH Aachen University, Germany.

Jörres, S. (2015). Aufbau und Inbetriebnahme eines Prüfstands für Hochleistungsverdampferrohre (in German). Bachelor's thesis, RWTH Aachen University, Germany.

Hackmann, J. (2015). Experimentelle Untersuchung von Glykol-Wasser-Gemischen als Arbeitsmittel in einer Adsorptionswärmepumpe (in German). Bachelor's thesis, RWTH Aachen University, Germany.

Bibliography

Aydin, D., Casey, S. P., and Riffat, S. (2015). The latest advancements on thermochemical heat storage systems. *Renewable and Sustainable Energy Reviews*, 41:356–367.

Baehr, H. D. and Stephan, K. (2011). *Heat and mass transfer.* Springer, Berlin, Heidelberg, 3rd edition.

Baehr, H. D. and Stephan, K. (2013). *Wärme- und Stoffübertragung (in German).* Springer Vieweg, Berlin, Heidelberg, 8th edition.

Bathen, D. and Breitbach, M. (2001). *Adsorptionstechnik (in German).* VDI-Buch. Springer, Berlin, Heidelberg, 1st edition.

Baumeister, J. and Weise, J. (2017). Evaporator and / or capacitor element with superficially embedded porous particles. Patent DE102016209082A1.

Bertossi, R., Romestant, C., Ayel, V., and Bertin, Y. (2012). Theoretical study and review on the operational limitations due to vapour flow in heat pipes. *Frontiers in Heat Pipes*, 3(2):1–9.

Birch, K. (2003). Estimating uncertainties in testing: an intermediate guide to estimating and reporting uncertainty of measurement in testing. British Measurement and Testing Association, Teddington, Middlesex, United Kingdom.

Braunschweig, N. and Paulussen, S. (2011). Evaporator for sorption machines, has heat exchanger provided with pipe or channel or combination of both, where fluid flows through pipe or channel. Patent DE102009053843A1.

Braunschweig, N., Paulussen, S., and Laufer, A. (2016). Surface feeding and distribution of a refrigerant for a heat exchanger in sorption machines. Patent EP2473811B1.

Burk, R., Wolff, T., Angermann, H.-H., Thiele, S., Schiehlen, T., Zwittig, E., Schroth, H., Felber, S., and Brunner, S. (2013). Module for a heat pump. Patent DE102011079586A1.

Burk, R. and Zwittig, E. (2010). Device for absorbing a fluid via capillary forces and method for manufacturing the device. Patent EP1918668B1.

Çengel, Y. A. and Boles, M. A. (2010). *Thermodynamics: an engineering approach.* McGraw-Hill, Boston, 7th edition.

Chase, M. W. (1998). *NIST-JANAF Thermochemical tables*, volume 9 of *Journal of Physical and Chemical Reference Data Monograph.* American Institute of Physics, Woodbury, NY, United States, 4th edition.

Chen, C. J., Wang, R. Z., Xia, Z. Z., Kiplagat, J. K., and Lu, Z. S. (2010). Study on a compact silica gel–water adsorption chiller without vacuum valves: design and experimental study. *Applied Energy*, 87(8):2673–2681.

Cho, H. J., Preston, D. J., Zhu, Y., and Wang, E. N. (2016). Nanoengineered materials for liquid–vapour phase-change heat transfer. *Nature Reviews Materials*, 2(2):16092.

Chu, K.-H., Enright, R., and Wang, E. N. (2012). Structured surfaces for enhanced pool boiling heat transfer. *Applied Physics Letters*, 100(24):241603.

Chua, H. T., Ng, K. C., Malek, A., Kashiwagi, T., Akisawa, A., and Saha, B. B. (1999). Modeling the performance of two-bed, sillica gel–water adsorption chillers. *International Journal of Refrigeration*, 22(3):194–204.

Cordray, D. R., Kaplan, L. R., Woyciesjes, P. M., and Kozak, T. F. (1996). Solid–liquid phase diagram for ethylene glycol + water. *Fluid Phase Equilibria*, 117(1–2):146–152.

Costa, A., Harm, M., Von Der Heydt, R., Högenauer-Lego, M., Kohlenbach, P., Kren, C., Lamp, P., Petersen, S., Schweigler, C., Storkenmaier, F., and Ziegler, F. (2006). Absorption chiller. Patent DE102004039327A1.

Cotter, T. P. (1965). Theory of heat pipes. Los Alamos Scientific Laboratory of the Universty of California, Los Alamos, NM, United States.

Cox, J. D., Wagman, D. D., and Medvedev, V. A. (1989). *CODATA Key values for thermodynamics.* Hemisphere Publishing Corp, New York.

Crößmann, F. (Juli 2016). *Untersuchung der Verdampfung aus strukturierten Oberflächen in Reinstoffatmosphäre (in German).* PhD thesis, TU Darmstadt, Darmstadt, Germany.

Curme, G. O. and Young, C. O. (1925). Ethylene glycol: a contribution of chemistry to the automobile antifreeze problem. *Industrial & Engineering Chemistry*, 17(11):1117–1120.

Dang, B. N., van Helden, W., and Luke, A. (2017). Investigation of water evaporation for closed sorption storage systems. *Energy Procedia*, 135:504–512.

de Gennes, P.-G., Brochard-Wyart, F., and Quéré, D. (2009). *Capillarity and wetting phenomena: drops, bubbles, pearls, waves*. Springer, New York, 1st edition.

Deutsches Institut für Normung e.V. (1990). Determination of values of surface roughness parameters R_a, R_z, R_{max} using electrical contact (stylus) instruments; terminology, measuring conditions. DIN 4768-1990-05.

Deutsches Institut für Normung e.V. (2009). Industrial platinum resistance thermometers and platinum temperature sensors. DIN EN 60751:2009-05.

Eral, H. B., 't Mannetje, D. J. C. M., and Oh, J. M. (2013). Contact angle hysteresis: a review of fundamentals and applications. *Colloid and Polymer Science*, 291(2):247–260.

Erdem, Ö. F. and Michel, D. (2005). ^{1}H MAS NMR investigations of ethylene glycol adsorbed in NaX. *The Journal of Physical Chemistry B*, 109(24):12054–12061.

Erdem, Ö. F., Tsuwi, J., Hong, S. B., and Michel, D. (2006). Molecular dynamics and glass-transition of ethylene glycol adsorbed in zeolites: influence of surface–molecule interactions, topology, and loading degree. *Microporous and Mesoporous Materials*, 94(1-3):261–268.

European Parliament (2014). Regulation (EU) No 517/2014 of the European Parliament and of the Council of 16 April 2014 on fluorinated greenhouse gases and repealing regulation (EC) No 842/2006.

Faghri, A. (1995). *Heat pipe science and technology*. Taylor & Francis, Washington, DC, 1st edition.

Faust, H. J. (2001). Evaporator for adsorption heat pump or refrigerator consisting of container for storage of liquid with heat supply or removal by lamination assembly immersed in liquid connected with heat carrying line. Patent DE10011688A1.

Fernandes, M. S., Brites, G., Costa, J. J., Gaspar, A. R., and Costa, V. (2014). Review and future trends of solar adsorption refrigeration systems. *Renewable and Sustainable Energy Reviews*, 39:102–123.

Fischer, S. (2015). *Experimental investigation of heat transfer during evaporation in the vicinity of moving three-phase contact lines.* PhD thesis, TU Darmstadt, Darmstadt, Germany.

Foster, P., Ramaswamy, V., Artaxo, P., Berntsen, T., Betts, R., Fahey, D. W., Haywood, J., Lean, J., Lowe, D. C., Myhre, G., Nganga, J., Prinn, R., Raga, G., Schulz, M., and Van Dorland, R. (2007). Changes in atmospheric constituents and in radiative forcing. In Solomon, S., Qin, D., Manning, M., Chen, Z., Marquis, M., Averyt, K., Tignor, M., and Miller, H., editors, *Climate change 2007: the physical science basis. Contribution of Working Group I to the fourth Assessment Report of the Intergovernmental Panel on Climate Change (IPCC)*, pages 129–234. Cambridge University Press, United Kingdom and New York, NY, USA.

Freni, A., Dawoud, B., Bonaccorsi, L., and Chmielewski, S. (2015). *Characterization of zeolite-based coatings for adsorption heat pumps.* SpringerBriefs in Applied Sciences and Technology. Springer International Publishing, Cham, Switzerland, 1st edition.

Friedrich, T. (2017). Refrigerant circulation. Patent DE102015220895A1.

Gao, L. and McCarthy, T. J. (2007). How Wenzel and Cassie were wrong. *Langmuir*, 23(7):3762–3765.

Gaugler, R. S. (1944). Heat transfer device. Patent US2350348A.

Giraud, F., Rullière, R., Toublanc, C., Clausse, M., and Bonjour, J. (2015). Experimental evidence of a new regime for boiling of water at subatmospheric pressure. *Experimental Thermal and Fluid Science*, 60:45–53.

Gorenflo, D., Goetz, J., and Bier, K. (1982). Vorschlag für eine Standard-Apparatur zur Messung des Wärmeübergangs beim Blasensieden (in German). *Wärme- und Stoffubertragung*, 16(2):69–78.

Graf, S., Redder, F., Bau, U., de Lange, M., Kapteijn, F., and Bardow, A. (2020). Toward optimal metal–organic frameworks for adsorption chillers: insights from the scale–up of MIL–101(Cr) and NH_2–MIL–125. *Energy Technology*, 8(1):1900617.

Graf, S. W. (2018). *A design approach for adsorption energy systems integrating dynamic modeling with small-scale experiments.* PhD thesis, RWTH Aachen University, Aachen, Germany.

Guildner, L. A., Johnson, D. P., and Jones, F. E. (1976). Vapor pressure of water at its triple point: highly accurate value. *Science*, 191(4233):1261.

Heinrich, C., Wittig, S., Albring, P., Richter, L., Safarik, M., Böhm, U., and Hantsch, A. (2015). *Sustainable cooling supply for building air conditioning and industry in Germany*, volume 05/2015 of *Climate Change*. Umweltbundesamt, Dessau-Roßlau.

Horstmann, S., Gardeler, H., Wilken, M., Fischer, K., and Gmehling, J. (2004). Isothermal vapor–liquid equilibrium and excess enthalpy data for the binary systems water + 1,2-ethanediol and propene + acetophenone. *Journal of Chemical & Engineering Data*, 49(6):1508–1511.

Hu, E. J. (1998). A study of thermal decomposition of methanol in solar powered adsorption refrigeration systems. *Solar Energy*, 62(5):325–329.

International Energy Agency (2018a). *World energy outlook 2018*. IEA, Paris.

International Energy Agency (2018b). The future of cooling: opportunities for energy-efficient air conditioning. *Technology report*.

International Institute of Refrigeration, Morlet, V., Dupont, J.-L., and Coulomb, D. (2017). The impact of the refrigeration sector on climate change. *Informatory Note on Refrigeration Technologies*, 35.

Jiao, Y. H. (2013). Review of materials for energy saving technology. *Applied Mechanics and Materials*, 275-277:2371–2374.

Joint Committee for Guides in Metrology (2008). Evaluation of measurement data - guide to the expression of uncertainty in measurement: JCGM 100:2008.

Joint Committee for Guides in Metrology (2009). Evaluation of measurement data - an introduction to the "Guide to the expression of uncertainty in measurement" and related documents: JCGM 104:2009.

Kell, G. S. (1975). Density, thermal expansivity, and compressibility of liquid water from 0 °C to 150 °C: correlations and tables for atmospheric pressure and saturation reviewed and expressed on 1968 temperature scale. *Journal of Chemical and Engineering Data*, 20(1):97–105.

Khan, M., Alam, K., Saha, B. B., Akisawa, A., and Kashiwagi, T. (2007). Study on a re-heat two-stage adsorption chiller – the influence of thermal capacitance ratio, overall thermal conductance ratio and adsorbent mass on system performance. *Applied Thermal Engineering*, 27(10):1677–1685.

Kim, H.-Y. and Kang, B. H. (2003). Effects of hydrophilic surface treatment on evaporation heat transfer at the outside wall of horizontal tubes. *Applied Thermal Engineering*, 23(4):449–458.

Kim, J. (2009). Review of nucleate pool boiling bubble heat transfer mechanisms. *International Journal of Multiphase Flow*, 35(12):1067–1076.

Köroğlu, B., Bogan, N., and Park, C. (2013). Effects of tube row on heat transfer and surface wetting of microscale porous-layer coated, horizontal-tube, falling-film evaporator. *Journal of Heat Transfer*, 135(4):041802.

Lang, R., Roth, M., Stricker, M., and Dawoud, B. (2001). Process for manufacturing an evaporator / condenser. Patent DE10053045A1.

Lang, R., Roth, M., Stricker, M., and Westerfeld, T. (1999). Development of a modular zeolite-water heat pump. *Heat and Mass Transfer*, 35(3):229–234.

Lanzerath, F. (2014). *Modellgestützte Entwicklung von Adsorptionswärmepumpen (in German)*. PhD thesis, RWTH Aachen University, Aachen, Germany.

Lanzerath, F., Bau, U., Seiler, J., and Bardow, A. (2015). Optimal design of adsorption chillers based on a validated dynamic object-oriented model. *Science and Technology for the Built Environment*, 21(3):248–257.

Lanzerath, F., Seiler, J., Erdogan, M., Schreiber, H., Steinhilber, M., and Bardow, A. (2016). The impact of filling level resolved: capillary-assisted evaporation of water for adsorption heat pumps. *Applied Thermal Engineering*, 102:513–519.

Lee, C. Y., Chun, T. H., and In, W. K. (2014). Effect of change in surface condition induced by oxidation on transient pool boiling heat transfer of vertical stainless steel and copper rodlets. *International Journal of Heat and Mass Transfer*, 79:397–407.

Lee, S., Köroğlu, B., and Park, C. (2012). Experimental investigation of capillary-assisted solution wetting and heat transfer using a micro-scale, porous-layer coating on horizontal-tube, falling-film heat exchanger. *International Journal of Refrigeration*, 35(4):1176–1187.

Lemmon, E. W., Huber, M. L., and McLinden, M. O. (2013). NIST Standard Reference Database 23: Reference Fluid Thermodynamic and Transport Properties-REFPROP, Version 9.1. National Standard Reference Data Series. National Institute of Standards and Technology, Gaithersburg, MD, United States.

Levin, M., Shaikh, F. Z., Demitroff, H. D., and Marshall, L. (2017). Adsorption air-conditioning system. Patent US9789746B2.

Li, T. X., Wang, R. Z., and Li, H. (2014). Progress in the development of solid–gas sorption refrigeration thermodynamic cycle driven by low-grade thermal energy. *Progress in Energy and Combustion Science*, 40:1–58.

Li, X. H., Hou, X. H., Zhang, X., and Yuan, Z. X. (2015). A review on development of adsorption cooling—novel beds and advanced cycles. *Energy Conversion and Management*, 94:221–232.

Lindstrom, P. J. and Mallard, W. G., editors (2013). *NIST Standard Reference Database Number 69*. Gaithersburg, MD, United States.

Liu, K., Vuckovac, M., Latikka, M., Huhtamäki, T., and Ras, R. H. A. (2019). Improving surface-wetting characterization. *Science*, 363(6432):1147–1148.

Liu, X., Wang, X., and Kapteijn, F. (2020). Water and metal-organic frameworks: from interaction toward utilization. *Chemical Reviews*, 120(16):8303–8377.

Masson-Delmotte, V., Zhai, P., Pörtner, H.-O., Roberts, D., Skea, J., Shukla, P., Pirani, A., Moufouma-Okia, W., Péan, C., Pidcock, R., Connors, S., Matthews, J., Chen, Y., Zhou, X., Gomis, M., Lonnoy, E., Maycock, T., Tignor, M., and Waterfield, T., editors (2018). *Global warming of 1.5 °C: an IPCC Special Report on the impacts of global warming of 1.5 °C above pre-industrial levels and related global greenhouse gas emission pathways, in the context of strengthening the global response to the threat of climate change, sustainable development, and efforts to eradicate poverty*. Special report. Intergovernmental Panel on Climate Change (IPCC), Geneva, Switzerland.

McGillis, W. R., Carey, V. P., Fitch, J. S., and Hamburgen, W. R. (1991). Pool boiling enhancement techniques for water at low pressure. In *Proceedings of the Seventh IEEE Semiconductor Thermal Measurement and Management Symposium*, pages 64–72.

McKay, I. S., Narayanan, S., and Wang, E. N. (2020). Bi-directional porous media phase change heat exchanger. Patent US20200217587.

Meunier, F. (2013). Adsorption heat powered heat pumps. *Applied Thermal Engineering*, 61(2):830–836.

Miles, D. J. and Shelton, S. V. (1996). Design and testing of a solid-sorption heat-pump system. *Applied Thermal Engineering*, 16(5):389–394.

Mittelbach, W. and Dassler, I. (2014). Method for executing an alternating evaporation and condensation process of a working medium. Patent EP2689201B1.

Mittelbach, W. and Wondra, F. (2009). Evaporator for use in e.g. heat pump, to partially or completely evaporating cooling water, has capillary structure provided in heating surface of heating element and is partially submerged in fluid reservoir. Patent DE102008028854A1.

Nath, A. and Bender, E. (1983). Isothermal vapor-liquid equilibriums of binary and ternary mixtures containing alcohol, alkanolamine, and water with a new static device. *Journal of Chemical & Engineering Data*, 28(4):370–375.

Nukiyama, S. (1966). The maximum and minimum values of the heat Q transmitted from metal to boiling water under atmospheric pressure. *International Journal of Heat and Mass Transfer*, 9(12):1419–1433.

Núñez, T. (2001). *Charakterisierung und Bewertung von Adsorbentien für Wärmetransformationsanwendungen (in German).* PhD thesis, Albert-Ludwigs-Universität, Freiburg, Germany.

Osborne, N. S., Stimson, H., and Ginnings, D. C. (1939). Measurements of heat capacity and heat of vaporization of water in the range 0° to 100°C. *Journal of Research of the National Bureau of Standards*, 23(2):197–260.

Plawsky, J. L., Fedorov, A. G., Garimella, S. V., Ma, H. B., Maroo, S. C., Chen, L., and Nam, Y. (2014). Nano- and microstructures for thin-film evaporation—a review. *Nanoscale and Microscale Thermophysical Engineering*, 18(3):251–269.

Potash, M. and Wayner, P. (1972). Evaporation from a two-dimensional extended meniscus. *International Journal of Heat and Mass Transfer*, 15(10):1851–1863.

Preston-Thomas, H. (1990). The international temperature scale of 1990 (ITS-90). *Metrologia*, 27(1):3–10.

pro KÜHLSOLE GmbH (2012). Glykosol N - datasheet: liquid coolant and heat carrier on basis of monoethylene glycol for technical applications.

Prüger, W. Z. (1940). Die Verdampfungsgeschwindigkeit der Flüssigkeiten (in German). *Zeitschrift für Physik*, 115(3-4):202–244.

Reay, D. A., Kew, P. A., and McGlen, R. J. (2014). *Heat pipes: theory, design and applications.* Butterworth-Heinemann (Elsevier), Amsterdam, 6th edition.

Restuccia, G., Freni, A., Russo, F., and Vasta, S. (2005). Experimental investigation of a solid adsorption chiller based on a heat exchanger coated with hydrophobic zeolite. *Applied Thermal Engineering*, 25(10):1419–1428.

Rezk, A., AL-Dadah, R., Mahmoud, S., and Elsayed, A. (2013). Investigation of ethanol/metal organic frameworks for low temperature adsorption cooling applications. *Applied Energy*, 112:1025–1031.

Rubitschek, F., Dupmeier, T., Reul, B., and Grewe, D. (2014). Evaporator tube for arrangement in an exhaust system and method for producing the evaporator tube with a porous sintered structure and steam channels. Patent DE102013103840A1.

Ruthven, D. M. (1984). *Principles of adsorption and adsorption processes*. A Wiley-Interscience Publication. John Wiley & Sons, New York, 1st edition.

Sabir, H. M. and Bwalya, A. C. (2002). Experimental study of capillary-assisted water evaporators for vapour-absorption systems. *Applied Energy*, 71(1):45–57.

Sabir, H. M. and ElHag, Y. (2007). A study of capillary-assisted evaporators. *Applied Thermal Engineering*, 27(8-9):1555–1564.

Sabir, H. M., ElHag, Y., and Benhadj-Djilali, R. (2008). Experimental study of capillary-assisted evaporators. *Energy and Buildings*, 40(3):399–407.

San, J.-Y. and Tsai, F.-K. (2014). Testing of a lab-scale four-bed adsorption heat pump. *Applied Thermal Engineering*, 70(1):274–281.

Schawe, D. (2001). *Theoretical and experimental investigations of an adsorption heat pump with heat transfer between two adsorbers*. PhD thesis, Universität Stuttgart, Stuttgart, Germany.

Schnabel, L. (2009). *Experimentelle und numerische Untersuchung der Adsorptionskinetik von Wasser an Adsorbens-Metallverbundstrukturen (in German)*. PhD thesis, TU Berlin, Berlin.

Schnabel, L., Scherr, C., and Weber, C. (2008). Water as refrigerant - experimental evaluation of boiling characteristics at low temperatures and pressures. In *International Sorption Heat Pump Conference (ISHPC 2008)*.

Schnabel, L., Volmer, R., Warlo, A., and Schossig, P. (2018). Evaluation of different evaporator designs and concepts for sorption processes using water as refrigerant. In *Proceedings of the 13th IIR-Gustav Lorentzen Conference on Natural Refrigerants*, pages 1403–1410, Paris. International Institute of Refrigeration.

Schnabel, L., Witte, K., Kowol, J., and Schossig, P. (2010). Vergleich von Verdampferstrukturen für das Kältemittel Wasser (in German). In *Deutsche Kälte-Klima-Tagung 2010 Magdeburg*. Deutscher Kälte- und Klimatechnischer Verein e.V. (DKV).

Schnabel, L., Witte, K., Kowol, J., and Schossig, P. (2011). Evaluation of different evaporator concepts for thermally driven sorption heat pumps and chillers. In *Proceedings of the International Sorption Heat Pump Conference (ISHPC 2011)*, pages 543–552.

Schnabel, L. and Witte, K. T. (2009). Water as refrigerant - evaporator development for cooling applications. In *5th Heat Powered Cycles Conference (HPC 2009)*.

Schulze, C. (2013). *A contribution to numerically efficient modelling of thermodynamic systems*. PhD thesis, TU Braunschweig, Braunschweig, Germany.

Schweikert, K., Sielaff, A., and Stephan, P. (2019). On the transition between contact line evaporation and microlayer evaporation during the dewetting of a superheated wall. *International Journal of Thermal Sciences*, 145:106025.

Schweizer, N., Stephan, P., and Winter, M. (2011). Evaporator for evaporation of liquid coolant, has housing which has inlet opening for liquid coolant and outlet opening for evaporated coolant. Patent DE102010016644A1.

Seiler, J., Lanzerath, F., Jansen, C., and Bardow, A. (2019). Only a wet tube is a good tube: understanding capillary-assisted thin-film evaporation of water for adsorption chillers. *Applied Thermal Engineering*, 147:571–578.

Sharafian, A. and Bahrami, M. (2014). Assessment of adsorber bed designs in waste-heat driven adsorption cooling systems for vehicle air conditioning and refrigeration. *Renewable and Sustainable Energy Reviews*, 30:440–451.

Sharma, V. K., Mitra, S., Maheshwari, P., Dutta, D., Pujari, P. K., and Mukhopadhyay, R. (2010). Effect of guest-host interaction on the dynamics of ethylene glycol in H-ZSM5 zeolite. *The European Physical Journal Special Topics*, 189(1):273–277.

Shirazy, M. R., Blais, S., and Fréchette, L. G. (2012). Mechanism of wettability transition in copper metal foams: from superhydrophilic to hydrophobic. *Applied Surface Science*, 258(17):6416–6424.

Sokolov, N., Tsygankova, L., and Zhavoronkov, N. (1971). The liquid - vapor equilibrium in the water - ethylene glycol and water - 1,2 - proylene glycol systems at various pressures. *Theoretical Foundations of Chemical Engineering*, 5:817–820.

Spahn, H.-J. and Szuder, T.-F. (2016). Evaporator heat exchanger. Patent DE102014223250A1.

Spahn, H.-J. and Szuder, T.-F. (2017). Heat exchanger for an evaporator. Patent DE102015213320A1.

Spahn, H.-J., Szuder, T.-F., and Salg, F. (2016). Evaporator. Patent DE102014224137A1.

Stimson, H. (1969). Some precise measurements of the vapor pressure of water in the range from 25 to 100 °C. *Journal of Research of the National Bureau of Standards - A. Physics and Chemistry*, 73A(5):493–496.

Thimmaiah, P. C., Sharafian, A., Huttema, W., Osterman, C., Ismail, A., Dhillon, A., and Bahrami, M. (2016). Performance of finned tubes used in low-pressure capillary-assisted evaporator of adsorption cooling system. *Applied Thermal Engineering*, 106:371–380.

Thimmaiah, P. C., Sharafian, A., Rouhani, M., Huttema, W., and Bahrami, M. (2017). Evaluation of low-pressure flooded evaporator performance for adsorption chillers. *Energy*, 122:144–158.

United Nations (2015). Adoption of the Paris Agreement. In *Report of the Conference of the Parties on its twenty-first session*. United Nation Framework Convention on Climate Change (UNFCCC).

van Carey, P. (2018). *Liquid vapor phase change phenomena: an introduction to the thermophysics of vaporization and condensation processes in heat transfer equipment*. Taylor & Francis, New York, 2nd edition.

Villamanan, M. A., Gonzalez, C., and van Ness, H. C. (1984). Excess thermodynamic properties for water/ethylene glycol. *Journal of Chemical & Engineering Data*, 29(4):427–429.

Volmer, R., Eckert, J., Füldner, G., and Schnabel, L. (2017). Evaporator development for adsorption heat transformation devices – influencing factors on non-stationary evaporation with tube-fin heat exchangers at sub-atmospheric pressure. *Renewable Energy*, 110:141–153.

Volmer, R., Eckert, J., and Schnabel, L. (18. - 20. November 2015). Analyse geometrischer Einflussfaktoren auf die Niederdruckverdampfung von Wasser an Lamellenwärmeübertragern (in German). In *DKV-Tagungsbericht 2015 Dresden*. Deutscher Kälte- und Klimatechnischer Verein e.V. (DKV).

Volmer, R., Schalling, T., and Schnabel, L. (19. - 21. November 2014). Vermessung des dynamischen Verdampfungs- und Kondensationsverhaltens an Wärmeübertragern für adsorptionstechnische Anlagen (in German). In *DKV-Tagungsbericht 2014 Düsseldorf*. Deutscher Kälte- und Klimatechnischer Verein e.V. (DKV).

Wagner, W. and Pruß, A. (2002). The IAPWS formulation 1995 for the thermodynamic properties of ordinary water substance for general and scientific use. *Journal of Physical and Chemical Reference Data*, 31(2):387–535.

Wang, C.-Y. and Tu, C.-J. (1988). Effects of non-condensable gas on laminar film condensation in a vertical tube. *International Journal of Heat and Mass Transfer*, 31(11):2339–2345.

Wang, D., Zhang, J., Tian, X., Liu, D., and Sumathy, K. (2014a). Progress in silica gel–water adsorption refrigeration technology. *Renewable and Sustainable Energy Reviews*, 30:85–104.

Wang, L. W., Wang, R. Z., and Oliveira, R. G. (2009). A review on adsorption working pairs for refrigeration. *Renewable and Sustainable Energy Reviews*, 13(3):518–534.

Wang, R., Wang, L., and Wu, J. (2014b). *Adsorption refrigeration technology: theory and application*. John Wiley & Sons, Singapore, 1st edition.

Wang, R. Z., Xia, Z. Z., Wang, L. W., Lu, Z. S., Li, S. L., Li, T. X., Wu, J. Y., and He, S. (2011). Heat transfer design in adsorption refrigeration systems for efficient use of low-grade thermal energy. *Energy*, 36(9):5425–5439.

Wang, X., Chua, H. T., and Ng, K. C. (2005). Experimental investigation of silica gel–water adsorption chillers with and without a passive heat recovery scheme. *International Journal of Refrigeration*, 28(5):756–765.

Warlo, A., Schnabel, L., Volmer, R., Füldner, G., and Weise, J. (2018). Air-conditioning machine. Patent WO2018033418A1.

Warlo, A., Seifarth, H., Volmer, R., and Schnabel, L. (2017). Evaluation of innovative heat exchanger designs and operation modes for the subatmospheric evaporation of water. In *International Sorption Heat Pump Conference (ISHPC 2017)*.

Weast, R. C. (1976). *Handbook of chemistry and physics: a ready-reference book of chemical and physical data*. Crc Press, Cleveland, Ohio, United States, 57th edition.

Weibel, J. A., Garimella, S. V., and North, M. T. (2010). Characterization of evaporation and boiling from sintered powder wicks fed by capillary action. *International Journal of Heat and Mass Transfer*, 53(19-20):4204–4215.

Witte, T. (2016). *Experimentelle Untersuchungen zum Sieden in Metallfaserstrukturen im Bereich niederer Drücke (in German)*. PhD thesis, TU Darmstadt, Darmstadt, Germany.

Xia, Z. Z., Yang, G. Z., and Wang, R. Z. (2008). Experimental investigation of capillary-assisted evaporation on the outside surface of horizontal tubes. *International Journal of Heat and Mass Transfer*, 51(15-16):4047–4054.

Yekta-fard, M. and Ponter, A. B. (1985). Surface treatment and its influence on contact angles of water drops residing on polymers and metals. *Physics and Chemistry of Liquids an International Journal*, 15(1):19–30.

Yuki, K., Fukushima, K., Takemura, A., and Suzuki, K. (2015). Wettability and evaporation enhancement for heat transport devices by high performance oxide layer. In *Proceedings of the 2015 International Conference on Electronic Packaging and iMAPS All Asia Conference (ICEP-IAAC)*, pages 601–604.

Zou, A., Chanana, A., Agrawal, A., Wayner, P. C., and Maroo, S. C. (2016). Steady state vapor bubble in pool boiling. *Scientific reports*, 6:20240.

Aachener Beiträge zur Technischen Thermodynamik

ABTT 1
Philip Voll
Automated Optimization-Based Synthesis of Distributed Energy Supply Systems
1. Auflage 2014
ISBN 978-3-86130-474-6

ABTT 2
Johannes Jung
Comparative Life Cycle Assessment of Industrial Multi-Product Processes
1. Auflage 2014
ISBN 978-3-86130-471-5

ABTT 3
Franz Lanzerath
Modellgestützte Entwicklung von Adsorptionswärmepumpen
1. Auflage 2014
ISBN 978-3-86130-472-2

ABTT 4
Thorsten Brands
Einfluss der Gemischzusammensetzung auf die Verbrennung im Diesel- und GCAI-Motor
1. Auflage 2014
ISBN 978-3-95886-006-3

ABTT 5
Dominique Dechambre
Efficient Measurement of Liquid-Liquid Equilibria using Automation and Optimal Experimental Design
1. Auflage 2016
ISBN 978-395886-077-3

ABTT 6
Niklas von der Aßen
From Life-Cycle Assesement towards life-Cycle Design of Carbon Dioxide Capture and Utilization
1. Auflage 2016
ISBN 978-3-95886-080-3

ABTT 7
Matthias Lampe
Integrated Process and Organic Rankine Cycle Working Fluid Design in the Continuous-Molecular Targeting Framework
1. Auflage 2016
ISBN 978-3-95886-086-5

ABTT 8
Thomas Hülser
Optische Untersuchung der Zündvorgänge und deren Auswirkung auf die Verbrennung in PKW-Motoren
1. Auflage 2016
ISBN 978-3-95886-090-2

Aachener Beiträge zur Technischen Thermodynamik

ABTT 9
Malte Döntgen
Reaction Models from Reactive Molecular Dynamics and High-Level Kinetics Predictions
1. Auflage 2016
ISBN 978-3-95886-156-5

ABTT 10
Heike Schreiber
Experiments and Validated Models for Adsorption Thermal Energy Storage in Industrial and Residential Application
1. Auflage 2017
ISBN 978-3-95886-178-7

ABTT 11
André Dirk Sternberg
System-Wide Perspective for Life Cycle Assesment of CO_2-based C1-Chemicals
1. Auflage 2017
ISBN 978-3-95886-193-0

ABTT 12
Uwe Bau
From Dynamic Simulation to Optimal Design and Control of Adsorption Energy Systems
1. Auflage 2018
ISBN 978-3-95886-216-6

ABTT 13
Christian Jens
Modellbasiertes Design von Produkt, Lösungsmittel und Prozess für die Ameisensäure-synthese aus CO_2 und H_2
1. Auflage 2018
ISBN 978-3-95886-231-9

ABTT 14
Jan David Scheffczyk
Integrated Computer-Aided Design of Molecules and Processes using COSMO-RS
1. Auflage 2018
ISBN 978-3-95886-236-4

ABTT 15
Björn Bahl
Optimization-Based Synthesis of Large-Scale Energy Systems by Time-Series Aggregation
1. Auflage 2018
ISBN 978-3-95886-210-1

Aachener Beiträge zur Technischen Thermodynamik

ABTT 16
Bastian Liebergesell
A Milliliter-Scale Setup for the Efficient Characterization of Multicomponent Vapor-Liquid Equilibria Using Raman Spectroscopy
1. Auflage 2018
ISBN 978-3-95886-247-0

ABTT 17
Stefan Wilhelm Graf
A Design Approach for Adsorption Energy Systems Integrating Dynamic Modeling with Small-Scale Experiments
1. Auflage 2018
ISBN 978-3-95886-258-6

ABTT 18
Sebastian Kaminski
Quantum-Mechanics-Based Prediction of SAFT Parameters for Non-Associating and Associating Molecules Containing Carbon, Hydrogen, Oxygen and Nitrogen
1. Auflage 2019
ISBN 978-3-95886-270-8

ABTT 19
Maike Renate Hennen
Decision Support for the Synthesis of Energy Systems by Analysis of the Near-Optimal Solution Space
1. Auflage 2019
ISBN 978-3-95886-277-7

ABTT 20
Peyman Yamin
COSMO-RS-Based Methods for Improved Modelling of Complex Chemical Systems
1. Auflage 2019
ISBN 978-3-95886-288-3

ABTT 21
Meltem Erdogan
Assessement of Adsorbents for Drying by Experiments and Dynamic Simulations
1. Auflage 2019
ISBN 978-3-95886-303-3

ABTT 22
Christian Schulz
SRS/LIF-Messungen zur Charakterisierung rußarmer dieselähnlicher Flammen von alternativen Kraftstoffen und n-Heptan
1. Auflage 2019
ISBN 978-3-91886-310-1

Aachener Beiträge zur Technischen Thermodynamik

ABTT 23
Peter Beumers
Physically-Based Models for the Analysis of Raman Spectra
1. Auflage 2019
ISBN 978-3-95886-319-4

ABTT 24
Arne Kätelhön
Technology Choice Model for Consequential Life Cycle Assessment
1. Auflage 2019
ISBN 978-3-95886-324-8

ABTT 25
Christine Peters
Measurement of Multicomponent Diffusion in Liquids Using Raman Microspectroscopy and Microfluidics
1. Auflage 2020
ISBN 978-3-95886-337-8

ABTT 26
Dinah Elena Hollermann
Reliable and Robust Optimal Design of Sustainable Energy Systems
1. Auflage 2020
ISBN 978-3-95886-346-0

ABTT 27
Thomas Raffius
Laserspektroskopische Analyse von selbstzündenden motorischen Einspritzstrahlen alternativer Biokraftstoffe
1. Auflage 2020
ISBN 978-3-95886-358-3

ABTT 28
Johannes Schilling
Integrated Thermo-Economic Design of Processes and Molecules Using PC-SAFT
1. Auflage 2020
ISBN 978-3-95886-368-2

ABTT 29
Nils Julius Baumgärtner
Optimization of Low-Carbon Energy Systems from Industrial to National Scale
1. Auflage 2020
ISBN 978-3-95886-385-9

ABTT 30
Ludger Wolff
From Model-based Experimental Design and Analysis of Diffusion and Liquid-Liquid Equilibria to Process Applications
1. Auflage 2020
ISBN 978-3-95886-402-3

Aachener Beiträge zur Technischen Thermodynamik

ABTT 31
Andrej Gibelhaus
A Model-based Framework for Optimal Systems Integration of Adsorption Chillers
1. Auflage 2021
ISBN 978-3-95886-406-1

ABTT 32
Jan Michael Seiler
Debottlenecking the Evaporator in Water-Based Adsorption Chillers
1. Auflage 2021
ISBN 978-3-95886-407-8